Tree Breeding

and

Improvement

Theories and Techniques

Related Books

Title	*Author*
• 21st Century Biology and Agriculture: Textbook Series (Vol. 1-5)	*A. Raman*
• Agroforestry for Climate Resilience and Rural Livelihood	*Inder Dev*
• Agroforestry Theory and Practices	A.J. Raj
• Biodiversity of Ranthambhore Tiger Reserve Rajasthan	*V. Singh*
• Bioinoculants for Sustainable Agriculture and Forestry	*S.M. Reddy*
• Compensatory Afforestation in Rajasthan: An Evaluation	*G. Singh*
• Ecosystem Valuation: Theory and Practice for Policy Making	*D.S. Kumar*
• Forest Ecology	*S.S..Sagwal*
• Forest Fires and their Control *E.S. Artsybashev*	
• Forest Mensuration: A Handbook for Practitioners	*G.J. Hamilton*
• Forest Nursery Manual:Production of Bareroot Seedlings	*M. L.Duryea*
• Forest Nursery: How to Raise and Manage	*S.S. Sagwal*
• Forest Tree Seeds: Handbook	*S.S. Sagwal*
• Forest Tribology and Anthropology	*V.M. Mhaiske*
• Forestry Extension Methods	*D. Sim*
• Forestry Technologies: A Complete Value Chain Approach	*K.T. Parthiban*
• Forestry: A Subjective Guide for IFS Aspirants	*K.T. Parthiban*
• Grassland Techniques and Management	L.N. Singh
• Handbook of Forest Tree Seeds**	*S. S. Sagwal*
• Industrial Agroforestry Perspectives and Prospectives	*K.T. Parthiban*
• Introduction to Forestry & Agroforestry	*K.T. Parthiban*
• Nursery and Plantation Practices in Forestry	*Vinod Kumar*
• Objective Forestry: For All Competitive Examination *3rd Ed***	*K.T. Parthiban*
• Operations Research	*P.K. Mohanty*
• Pests of Forest Importance and their Management	*B.K. Tyagi*
• Plant Nursery Management: How to Start and Operate A Plant Nursery	*P. K. Ray*
• Plantation and Agroforestry Pulpwood Value Chain Approach	*K.T. Parthiban*
• Practical Manual on Plantation Forestry*Pankaj Panwar*	
• Rajasthan Ke Pavitra Upavan: Sambhavit Khatein Tatha Prabhandh Nitiyan (Hindi)	*G. Singh*
• Wildlife Ecology and Conservation (21st Century Biology and Agriculture: Textbook Series)	*M. Balakrishnan*

Tree Breeding and Improvement

Theories and Techniques

Dr. K.T. Parthiban
Dr. N. Krishnakumar
Dr. P.S. Devanand

FOREST COLLEGE AND RESEARCH INSTITUTE
TAMIL NADU AGRICULTURAL UNIVERSITY
METTUPALAYAM - 641301

SCIENTIFIC PUBLISHERS

Published by:

SCIENTIFIC PUBLISHERS (INDIA)
5A, New Pali Road, P.O. Box 91
Jodhpur 342 001, India
E-mail: info@scientificpub.com
Website: www.scientificpub.com

Print: 2025

ISBN: 978-93-89412-83-3
eISBN: 978-93-89412-84-0

Printed in India

About the Authors

Dr. K.T. Parthiban is borne at the faculty of Forestry and is currently positioned as Dean, Forest College and Research Institute, TNAU, Mettupalayam. He is actively involved in teaching, research and extension activities for the past 25 years. He has conceived the value chain models for Industrial Agroforestry and promoted in association with wood based industries. He is also pioneer in development of varieties in trees. He has handled numerous projects funded by DST, DBT, ICAR, NOVOD Board and several wood based industries. He has published 12 books and over 100 research papers. Some of his books like Objective Forestry, Subjective Forestry to IFS Aspirants and Forestry Technologies have profound utility by the students, professional foresters and wood based industries. Dr. K. T. Parthiban travelled widely in European, South American and Asian countries and contributed towards Forestry Research and Development.

Dr. N. Krishnakumar is borne at the faculty of Forestry and is currently working as a Forest Range Officer (Trainee) in the Tamil Nadu Forest Department. He is specialized in tree breeding and involved in varietal development programme of Thornless bamboos and Sandalwood. He is a recipient of Rajiv Gandhi National Fellowship and has published 3 books and several research papers in the journals of national and international repute.

Dr. P.S. Devanand is borne at the faculty of Agriculture and involved in genetic improvement of Annuals and Perennials. Dr. P.S. Devanand has obtained his Doctorate from Kyoto University, Japan and Post Doc from University of California, Riverside, USA. He is actively involved in plant genetic resources conservation and sustainable utilization through traditional and molecular breeding approaches. He has more than 25 international research papers to his credits.

TAMIL NADU AGRICULTURAL UNIVERSITY

Prof. N. KUMAR, Ph.D., F.H.S.I.,
Vice-Chancellor

Coimbatore - 641 003
Tamil Nadu, India

FOREWORD

Tree breeding has gained potential attraction across the country due to growing wood demand coupled with restriction on wood supply from natural forest. The National Forest Policy (1988) and Agroforestry Policy (2014) have directed for increased participation of wood based industries in agroforestry towards meeting their raw material resources. It has attracted increased attention in tree breeding as well to evolve new varieties and clones for higher productivity, improved wood quality and the associated resistance towards various stresses. The growing concern on depletion of forest genetic resources, genetic erosion, over exploitation etc. have also indicated tree breeding as a conservation tool.

Keeping this in view, ICAR has recommended starting an exclusive PG programme in Tree Breeding in all SAUs accredited with B.Sc and M.Sc programme in Forestry. Accordingly, several institutions have started an exclusive Tree Breeding programme at PG level and created specialization in doctoral studies. This research and educational activities have demanded wide range of text books incorporating the fundamental and application aspects of Tree Breeding and Improvement. In India, over 25 different tree species have been incorporated in tree breeding and there is no reference material for use in tree breeding both for academic and practical applications. Above all, several competitive examinations incorporated the fundamental and application aspects of tree breeding both through objective and subjective modes. The existing books on Tree Breeding are based on experiences with temperate species and as on date, there is no exclusive text book to deal with practical aspects of tropical tree breeding.

Under such circumstances, I am happy to see that the scientific team of FCRI has brought out a comprehensive text book entitled **"Tree Breeding and Improvement-Theories and Techniques"** which has 19 chapters encompassing all facets of tree breeding which will cater to the needs of UG, PG and Ph. D programme in Forestry and Tree Breeding. Above all, the book will act as a ready reckoner for all academic and competitive examinations. The tree breeding scientists involved in varietal development will find this book as practical guide for them towards improving and conserving the Forest Genetic Resources.

I congratulate the authors for their efforts in bringing out such a basic text book which is being demanded by students, scientists and professional foresters.

I wish the authors every success.

Date: 17.03.2020
Place: Coimbatore

(N.KUMAR)

• Tel : Off.: +91 422 2431788 *Res. : +91 422 2430887 *Fax: +91 422 2431672 *Email: vctnau@tnau.ac.in

Contents

1

BASIC AND COMPONENTS OF TREE IMPROVEMENT

INTRODUCTION

Forests in India have played a significant and critical role in meeting the domestic industrial raw material requirements besides catering to the needs of socio economic and livelihood opportunities of the people. These forests were traditionally managed with the intensive silvicultural system without taking into consideration the breeding value of the forest genetic resources. This has witnessed the low productivity ($0.5m^3$/ ha/ annum) of the existing natural forests compared to the global average of $2.1m^3$/ ha/ annum. The growing population, urbanization, industrialization and the associated science and technology development in the country have demanded increased use of wood and wood products. Till the recent past, these wood and wood products demand has predominantly been met from the natural forests; but due to the policy and legal restrictions, the supply from natural forest has decreased very rapidly. The Forest Conservation Act 1980 and the subsequent directions of the Honorable Supreme Court in the year 1996, have almost totally stopped the felling operation in natural forests and reduced the supply of wood and wood products to the wood based industries.

The growing wood demand, restriction on felling coupled with poor productivity of India's forest have ushered in a total mismatch between demand and supply of both domestic and industrial wood requirement. These developments have witnessed large scale promotion of forestry plantation outside its natural range particularly through agro forestry and farm forestry systems. These promotional activities compelled the need for the improvement of Forest Genetic Resources through systematic tree breeding and improvement approaches. Traditional tree breeding and improvement works were concentrated only on limited species like teak, sal, deodar and few pines. However due to increasing demand for wide range of wood and wood products' over 25 different trees species have been prioritized for varied industrial utility and demands intensive tree breeding and improvement approach to increase the productivity, reduce the rotation, augment the wood quality, early flower induction, development of varieties for stress tolerance and other specific objectives. As of now, there is no comprehensive text book is available to cater to the academic needs of forestry colleges in the country and also for practically applied breeding methods for adoption by researchers and professional foresters. Against this backdrop the current book on "Tree Breeding and Improvement - Theories and Techniques" has been conceived and presented for use by stakeholders.

HISTORICAL DEVELOPMENTS IN TREE BREEDING

Forest tree breeding and improvement program compared to agriculture is only an young science which had its recognizable start eight to nine decades back. Globally, Sweden was the pioneer in tree breeding and other countries subsequently initiated and strengthened the tree breeding program which resulted in productivity improvement. Predominantly, the tree breeding program was approached through selection and hybridization methods with little initiatives in mutation breeding. The tree breeding and improvement program attempted globally and regionally has been traced and presented.

The global forest genetists and breeders as traced and presented in Zobel and Talbert, (1984) is furnished below for understanding the tree breeding developments across the world.

1717	Bradley (England)	Importance of seed origin
1760	Duhamel de Monceau (France)	Inheritance : oak
1761	Koehlreuter (Germany)	Hybridization
1787	Bursdorf (Germany)	Plantation for seed production

1840	Marrier de Boisdhyver (France)	Vegetative Propagation
1840	de Vilmorin (France)	Fir hybrids
1845	Klotzsch (Berlin)	Intraspecific hybrids: oak, elm, and alder
1904	Cieslar (Austria)	Provenances : Larch and oak
1905	Engler (Switzerland)	Elevation differences of species: fir, pine, spruce, larch and maple
1905	Dengler (Germany)	Provenance tests: fir and spruce
1906	Andersson (Sweden)	Vegetative propagation
1907	Sudworth, Pinchot (U.S.A)	Breeding nut and other forest trees
1908	Oppermann (Denmark)	Straightness : beech and oak
1909	Johannsen (Sweden)	Elite stands
1909	Sylven (Sweden)	Self-Pollination : Norway spruce
1912	Zederbauer (Austria)	Crown form: Austrian pine
1918	Sylven (Sweden)	Seed orchards
1922	Fabricius (Austria)	Plantation for seed production
1923	Oppermann (Denmark)	Seedling seed orchards
1924	Schreiner (U.S.A)	Poplar breeding
1928	Burger (Switzerland)	Pine selection
1928	Bates (U.S.A)	Seed orchards
1930	Larsen S. (Denmark)	Controlled Pollination : larch
1930	Heikinheimo and M. Larsen (Finland)	Curly grain : birch
1930	Nilsson – Ehle, Sylven, Johnsson Linquist (Sweden)	Pine and aspen breeding
1935	Nilsson – Ehle (Seden)	Triploid aspen

In India the tree breeding is very young and organized tree breeding programmes started only after 1960 with emphasis on only few species. However after 1990s several wood based industries, research institutions and organizations spread across the country initiated tree breeding and improvement program in almost over 25 different tree species which are of commercial importance and amenable for industrial agroforestry. The significant developments in tree breeding and improvement programme in India is furnished.

Year	Developments	Source
1950	Systematic introduction of European and euro American popular made by Uttar Pradesh forest Department	Dinesh Kumar and Singh, 2012
1959 - 1976	FRI introduced 199 Poplar clones	Kaul and Sharma, 1982
1977 - 1981	Another set of 132 poplar clones introduced by FRI Wimco also introduced exotic clones of poplars directly from abroad	
1956	Karyotyping studies in pines	Mehra and Khoshoo 1956 Kumar *et al.*, 1966 Upathaya and Kedharnath, 1970
1960	Organized breeding program in forest trees initiated by forest research institute, Dehradun	Kedharnath, 1984
1960	Species like Teak (*Tectona grandis*), Chir pine (*Pinus roxburghii*), Semul (*Bombax ceiba*) and eucalyptus were prioritized	
1961	Detailed forest genetics and tree breeding programme presented at the 10 the silvicultural conference held at Dehradun	Kedharnath and Raizada, 1961
1963	Semal Improvement programme started	Kedharnath and Venkatesh, 1963
1968	700 plus trees in teak were identified and deployed for establishing clonal seed orchard using budding and cleft grafting techniques	Rawat and Kedharnath, 1968

Year	Developments	Source
1970	Establishment of first clonal seed orchard at New forests Dehra Dun	Kedharnath *et al.*, 1970
1975	Comparitive testing of *Tectona grandis and Tectona hamiltoniana* for resistance towards *Eutectona machaeralis*	Kedharnath and Pratap Singh, 1975
1977	Hybrid evaluation in eucalyptus	Venkatesh and Sharma, 1977
1979	Teak improvement programme in Tamil Nadu	Kumaravelu, 1979
1979	Tree improvement in pines attempted	Khosla *et al.*, 1979 Uniyal and Thapliyal, 1979
1979	Poplar improvement programme	Kedharnath, 1979
1980	Development of tissue culture protocol for major tree crop by National Chemical laboratory	Gupta *et al.*, 1980
1981	Meristem culture in *Eucalyptus citriodora*	Gupta *et al.*, 1981
1982	Improvement programme in *Populus ciliata*	Khurana and Khosla, 1982
1984	Provenance variation in 15 species of Eucalyptus for cylinderocladium blight disease	Jayasri *et al.*, 1984
1984	Meristem culture in *E. grandis*	Lakshmi sita *et al.*, 1984
1984	Variability studies on eucalyptus	Krishnasamy *et al.*, 1984
1990	Dr Y. S. Parmer University of Horticulture and Forestry introduced open pollinated families of *Populus deltoides*	Dinesh Kumar and Singh, 2012
1980 -as on date	Forest College and Research Institute initiated tree breeding programmes for over 15 different tree species	Surendran *et al.*, 1992 Parthiban *et al.*, 2015

Year	Developments	Source
1990 - 2000	ITC made significant improvement in eucalyptus productivity through organized clonal developmental programme and released several eucalyptus clones on their own	Kulkarni, 2000
2008	First variety in casuarina - MTP CA 1 was released by TNAU	TNAU, 2008
2009	IFGTB released casurina and eucalyptus clones	IFGTB, 2009
2010	TNAU released a hybrid casuarina clone MTP 2 with a productive potential of 150 tonnes / ha in three years	Parthiban *et al.*, 2010
2012	TNAU released first *Melia dubia* variety in India Melia MTP 1	Parthiban *et al.*, 2012
2013	TNAU released eucalyptus clone with higher pulp yield	TNAU, 2014
2018	TNAU released second variety in *Melia dubia* variety in India Melia MTP 2	Parthiban *et al.*, 2019
2019	TNAU released first clonal variety in Kadam	TNAU, 2019

DEFINITION OF TERMINOLOGIES

Tree breeding : It is an art and science of improving the genetic makeup of tree species like relation to their economic utility (or) usage.

Tree Improvement : It is a tool of improvement the genetic makeup of the trees through the combined action of breeding methods and silvicultural management.

Forest Genetics : The study of heredity in forest trees.

Forest Genetic Resources : Forest genetic resources (FGR) are the heritable materials maintained within and among tree and other woody plant species that are of actual or potential economic, environmental, scientific or societal value.

CONCEPT OF TREE BREEDING AND IMPROVEMENT

The improvement of any species relies on the following basic factors.

i) The availability of base or breeding population like provenances or geographic sources, seed sources, progeny, clones etc.

ii) The determination of amount, kind and causes of variability within a species.

iii) Development of varieties and species with improved quality and desired yield.

iv) Mass multiplication of developed or improved species and varieties for commercial application.

v) Collection, maintenance and evaluation of broader forest genetic resources for future and advance tree improvement programme.

OBJECTIVES OF TREE BREEDING

The Primary objective of any tree breeding is to develop superior genetic stocks with the potential of high yield and better wood quality. However, the breeding objective differs from species to species based on the utility and adoptability. The range of tree breeding objectives currently in practice is furnished below.

1. Development of High Yielding Varieties

The major objectives of any tree breeding program is the development of varieties with high yielding capacity in term of timber, pulp wood, ply wood, matchwood, biomass, biofuel etc., depending on the consumer industry requirement. The productivity is aimed at to improve the levels from $10m^3$/ ha/ annum to a maximum of over $25m^3$/ ha/ annum. This improvement in wood yield is pertaining to timber species which may yield solid wood for all wood based industries. In case of non-timber forest productivity, the breeding objective and improvement is entirely different which may be yield improvement for fruit, fodder, fiber, fuel, oil, tuber, bark, seed etc,. In a holistic perspective, depending on the species, tree breeding aims to improve wood yield or other minor forest product yield.

The following are the major species developed by TNAU for yield improvement.

S.No.	Species	Breeding Method	Yield
1.	Casuarina MTP 1	Clonal Selection	100 tonnes / ha
2.	Casuarina MTP 2	Hybrid Development	150 tonnes / ha
3.	Eucalyptus	Clonal Selection	130-150 tonnes / ha
4.	*Melia dubia* MTP 1	Clonal Selection	150 tonnes / ha
5.	*Melia dubia* MTP 2	Clonal Selection	200 tonnes / ha
6.	Kadam MTP 1	Clonal Selection	135-150 tonnes / ha
7.	Teak (FCRITK01)	Progeny Selection	250 tonnes / ha
8.	*Gmelina arborea* (FCRI SS)	Seed source	200 tonnes / ha
9.	*Ailanthus excelsa* (MTP AE 17)	Progeny Selection	200 tonnes / ha
10.	*Dalbergia sissoo* (DS 18)	Clonal Selection	150 tonnes / ha

2. Improvement of wood quality

Breeding objective also aims to improve the wood quality which differs from species to species and within a species it varies from base to top upwards. The improvement of wood quality will involve predominantly, wood density (timber quality), cellulose content, hemi-cellulose contents and Lignin (chemical quality), Calorific value, ash content (energy value), low bark content (higher wood yield) etc, are the major quality parameters intended for breeding a variety is a species. Similarly in case of species amenable for Non Timber Forest Products, the objective may be based on the utility. For example in tree borne oilseeds, breeding objective may be improving both oil content and quality. In case of gum and resin, the objective is to improve the yield with superior chemical quality. The following are the variety developed by TNAU for improved wood quality

S.No	Species	Variety	Major wood quality improved
1.	Eucalyptus	EC 48 or MTP 1	High pulp yield (48% and above)
2.	Melia	MTP 2	Pulp yield - 15.5% Kappa number - 20
3.	Jatropha	FCRI HC 21	Oil content (33%)
4.	Subabul	FCRI LL 15	44% pulp yield
5.	Kadam	MTP 1	High fissility
6.	Melia	MTP 1 & MTP 2	High veneer recovery

3. Development of biotic resistant variety

Forests in general and trees in particular are predominantly free from pest and disease and there is no biological threat for these species. However the intensive promotion of Agroforestry and plantation forestry activities, have witnessed large scale promotion of single species through establishment of monocultures. The classical examples are promotion of poplars in Northern India, Teak, Eucalyptus, Casuarina, Subabul, Ailanthus, *Dalbergia sissoo* and Melia in Southern India.

These monocultured plantation witnessed several damages due to pest and diseases and incur significant yield and economical loss to the tree growing farmers and associated wood based industries. Some of the pest due to their inherent habitation in the wood logs cause significant wood quality damage. The following are the major epidemic pest and diseases which warranted developing of resistance.

S.No.	Species	Major pest and diseases	Casual Organisms
1.	Teak	Skeletonizer	*Hyblaea puera*
2.	Ailanthus	Defoliator	*Eligma narcissus*
3.	Eucalyptus	Gall wasp	*Leptocybe invasa*
4.	Casuarina	Stem wilt	*Trichosporium vesiculosum*
5.	Melia	Root rot	*Ganoderma lucidum*

4. Development of Abiotic resistant varieties

Tree species also suffer due to abiotic stress like drought, soil salinity, alkalinity, high temperature, cold, frost and wind, which result in significant economic loss. The classical example is dying back in sissoo in northern India which caused destructive damage to sissoo plantation. Similarly the recent climate change also witnessed changes in reproductive system of several tree crops. All these damages due to abiotic stress demand development of resistant varieties in trees also.

5. Development of short duration tree varieties

Trees are generally considered as a long duration crop which in most cases detracts the attraction of farmers and industries. In India, most farmers lands are small scale land holders and the farmers demand short duration tree crops. Hence one of the major objectives of breeding is to develop short duration varieties which could be amenable for harvest between one and five years, of plantation establishment. The FCRI, Mettupalayam has developed several short duration varieties and promoted through consortium of industrial agroforestry approach.

S.No.	Species	Variety	Rotation/Duration
1.	*Casuarina*	MTP - 1	36 months
2.	*Casuarina*	MTP - 2	30 months
3.	*Melia*	MTP - 1	60 months for ply wood 36 months for pulp wood
4.	*Melia*	MTP - 2	18 months for pulp wood 48 months for ply wood
5.	*Subabul*	FCRI, LL15	30 months for pulp wood 8 years for timber
6.	*Kadam*	MTP - 1	3 years pulp wood 5 years ply wood

6. Development of photo and theromo insentive varieties

The major objective of tree breeding is also to develop tree varieties which are insensitive to light and temperature which arises due to gravity climate change associated risk and uncertainties, Development of climate variable based insensitive varieties will be helpful to cross the natural habitation area (or) cultivation boundaries of varieties. If such varieties are developed, there are chance of introducing them to extremes of weather conditions line hot and cold deserts.

Example: Tropical species can be introduced to temperature climate and vice versa.

7. Developing of tree varieties with synchronous flowering / maturity

Several tree species particularly tree borne oil seeds (TBOs) like Neem, Pungam, Mahua, Calophyllum and Jatropha exhibit wide variation in flowering and become a major hindrance for the collection of fruits. Since these trees exhibits continuous flowering for a period of > 6 months and demand more labourers for collection and the increase cost and time. Under such circumstance the breeding objective may be to develop a variety with a synchronous maturity which will facilitate easy monitoring, collection and will reduce the cost and time.

8. Developing of varieties with desirable tree characters

The breeding objectives are rapidly expanding to cater to the growing needs of domestic and industrial sector. It involves, the developing or designing a tree with a straight cylindrical bole, apical dominance resistance to abiotic and biotic stress, low branching character, self pruning ability etc. These tree characteristic may be of great use to the tree growing farmers and also to the consuming industries.

Example:

- Clean bole development for Melia for face and core veneer
- Clean bole development in Ailanthus for match splints
- Self pruning ability in *Melia dubia*
- Smooth and low bark content in Eucalyptus
- Low branches in teak and red sanders

9. Development of varieties for diverse environmental adoptability

The adoptability refers to the suitability of forestry species for commercial cultivation over a wide range of environmental conditions, like extreme cold and hot, heavy wind, shifting, sand dunes, deserts etc,. The tree breeding should inculcate these factors into account to develop a variety amenable for wide environment range of adaptation.

10. Development of toxin free variety

In recent times, the objective has been expanded to develop tree varieties free from toxic compounds particularly in those varieties which are amenable for human and animal consumption.

Example:

- Mimosin present in *subabul* is toxic to cattle and hence removal of this anti nutritional factor is one of the major objectives of the breeder.
- Presence of phorbol ester in *jatropha* prohibits the use of oil cake for animal feed. In this case, the objective of the breeding is to develop non-toxic varieties in *jatropha* .

Hence the breeding objective should include detoxification as the major objective particularly for the trees amenable for fruit, food and fodder production.

COMPONENTS OF TREE IMPROVEMENT

The tree breeding and improvement basically involve three major components taking into consideration the (i) Biological resources, (ii) Breeding methods deployed and the (iii) Means of multiplication of genetic resources.

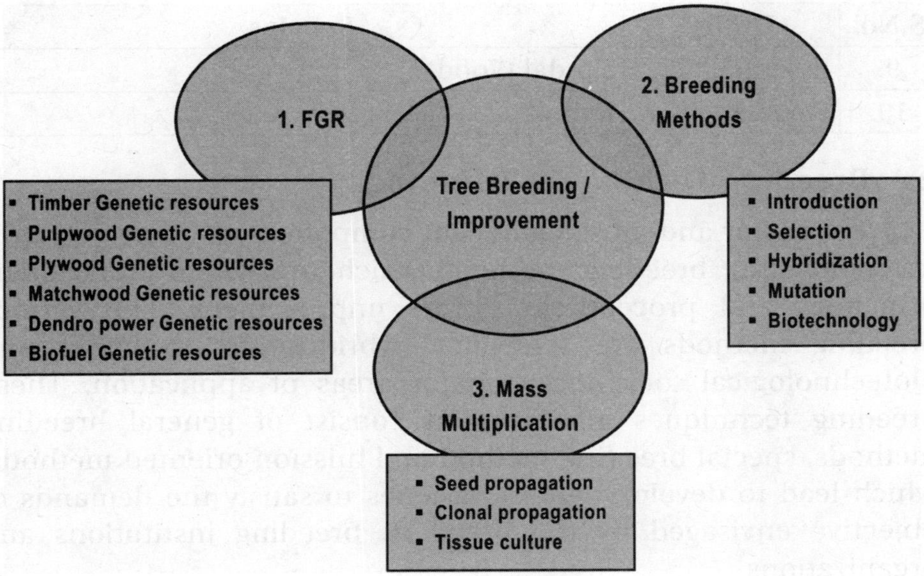

1. Forest Genetic Resources (FGR)

Refer to the base and breeding population deployed in tree breeding and improvement programme which represents more significance area of tree breeding and improvement. This FGR deals with collection, establishment evaluation, characterization and future utilization of biological resources both for domestic and industrial utility. This FGR will also act as a germplasm and satisfy the demands of existing conservation of FGR. The FCRI has established FGR park for the following species.

S.No.	Utility	Species
1.	Timber	Teak, Red sanders, Albizia, Silver oak, *Dalbergia sissoo*
2.	Plywood	*Eucalyptus, Melia, Toona, Acrocarpus, Swietenia*
3.	Pulpwood	Casuarinas, *Eucalyptus,* Subabul, bamboos, Lannea
4.	Match wood	*Ailanthus, Kadam, Sterculia*
5.	Bio fuel	Neem, Jatropha, Pungam, Punnai, Mahua
6.	Bioenergy	Prosopis, Acacia
7.	Dye	Bixa, Wrightia
8.	Silk	Mulberry

S.No.	Utility	Species
9.	Essential oil	Sandal Wood
10.	Flosses	Kapok

2. Breeding Techniques / Methods

The major and most significant component of tree breeding is to deploy tree breeding methods which involve several genetic principles and procedures in tree improvement. The various breeding methods are selection, hybridization, mutation and biotechnological tools forms major areas of application. These breeding techniques and method consist of general breeding methods, special breeding method and mission oriented methods, which lead to development of varieties to satisfy the demands of objective envisaged by the breeders, breeding institutions and organizations.

i. General breeding methods

It includes introduction, selection and hybridation methods towards regular developing of variety in tree crops.

ii. Special breeding methods

It incorporates special breeding method like mutation breeding, polyploidy breeding, distance breeding and molecular breeding, including genetic engineering methods.

iii. Mission mode breeding program

In certain times, to cater to the needs of urgent issue and to resolve the emergency problem in forestry sector, the mission mode approach like gall wasp resistance in eucalyptus need to be developed.

3. Mass multiplication

The yet another significance component of tree breeding is to produce Quality Planting Material (QPM) of the developed variety to cater to the needs of tree growing farmers, wood based industries, Afforestation and reforestation requirement of state forest departments and other user agencies. Mass multiplication is planned through two approaches, like seed based propagation and clone based propagation including micropropagation.

BENEFITS OF TREE BREEDING / IMPROVEMENT

- Tree breeding helps to assemble all genetic resources in the form of base and breeding population which may act as an *ex situ* conservation parks and facilitates domestication and conservation of Forest Genetic Resources.
- Tree breeding helps to evolve tree varieties with varied density levels (Low, Medium and High) which will benefit timber industries to deploy them in their product developmental activities.
- Tree breeding helps to develop High Yielding Short Rotation Varieties (HYSR), which will have wider application and adoption both by the tree growing farmers and wood based industries.
- Tree breeding has an advantage of inventory, identification and development of varieties for biotic resistance which will be having greater utility especially during the epidemic status of pest and diseases during the promotion of single species based monoculture plantation.
- The inventory of new and alternate species amenable for varied industrial utility will help to inculculate diverse species in all utilization programme. This will also help to eliminate the effects due to monoculture of single species based plantation development.
- Tree breeding mandates to evolve varieties amenable for wider environmental adaptation which will have a significant advantage of deploying these varieties in difficult sites. The existing deserts of both hot and cold ecosystem, coastal tracts of the whole country, roads and railway lands etc., will have a positive side of utilization, due to this breeding program
- Most wood based industries, particularly timber, plywood, MDF, Match splints, sports goods, agricultural implements, dendro power, biofuel etc., will be benefited through the tree breeding activities in order to get superior genetic stocks coupled with new and alternate species amenable for incorporation.
- The development of new species and varieties through tree breeding will help to protect the natural reserve forests as a

source of conservation and newly bred varieties will be deployed for commercial utilization.

- Tree breeding methods will be of greater use to protect and conserve the rare, endangered and threatened species (RET Species), besides helping them towards rehabilitation not only in their original habitats but also in the farm lands.
- Tree breeding also helps to evolve new varieties with alternated chromosome number through mutation breeding which will help to improve the wood quality of several species.
- Several high value species like sandal and redsander are nearly endangered and are threatened with extinction. If not protected, these species will be totally eliminated from the ecosystem. For such species, tree breeding and improvement programme will help to protect high value trees through conservation breeding and thereby help to conserve the high value and threatened genetic resources of the country.
- Above all, the tree breeding and improvement activities help to exchange Plant Genetic Resources for domestication, commercialization, utilization and conservation. This will also help to introduce several fast growing exotics and will satisfy the wood requirement of both domestic and industrial utility.

LIMITATION OF TREE BREEDING

- Trees by nature are large biological components which demand necessary care and maintenance which is a tedious process in all tree breeding activities.
- Most tree breeding programme have to be carried on a long term basis which demands sustained efforts of a breeder to follow the breeding process. But the administrative procedures may limit the participation of scientist continuously on the long term breeding process.
- Several tree breeding programme due to their long term nature, involve huge cost and expenditure and demand continuous funding support which may be difficult for several species.
- The reproductive mechanism of most trees will take several years and hybridization programme depends on the flowering which is one of the major limitation for the tree breeder and the tree breeding programme.

SUMMARY

Tree breeding and improvement is gaining significant attention and attraction among researchers, wood based industries and other professionals in order to develop superior genetic stocks to cater to the demands of domestic and industrial utility. The objectives of tree breeding are also expanding from the level of productivity improvement to the level of quality improvement, synchronised maturity, detoxification, earliness etc., which attracts several user agencies. This tree breeding involves three major components like FGR, Breeding methods and mass multiplication techniques. This chapter elaborates tree breeding value in forestry along with the approaches, advantages and limitations of tree breeding which may be of great reference to the graduates, scholars, professionals, researchers and practising foresters.

2

CENTER'S OF ORIGIN

INTRODUCTION

Each and every plant genetic resources has its own birth place on the earth and it differs from species to species. The birth place is the origin of crop plants on the earth and is pronounced as the Centers of Origin. For the particular crop, Center of origin is geographic area where the particular group of organisms (either domesticated or wild) originate first on the earth. This information on the origin of plants is very important particularly in plant breeding systems in order to identify and locate wild relatives, related species and new genes. The information on Center of origin and the associated knowledge on the origin of plants are very important in order to conserve the genetic resources and to avoid genetic erosion. The knowledge on origin of plant system is also helpful to avoid the loss of germplasm due to the loss of ecotypes and habitat. Under such circumstances, the information on Center of origin receives significant attention and will act as a great source to the plant breeders to conserve the existing genetic stocks and to develop new varieties to cater to the needs of the society.

DEFINITION OF TERMINOLOGIES

Center of origin : It is the geographic area wherein the particular group of organism which may be domesticated or wild originated first on the earth.

Center of diversity : Center of diversity is the place wherein the species is likely to express the greater variation in a region with a varied climatic and other ecological condition.

Gene Center	:	It is an area of high diversity and it must represent the Center of origin.
Genetic erosion	:	The loss of genetic resources due to over exploitation, monoculture, promotion of high yielding varieties and changes in land use system.
Monocentric crop	:	They are relatively new crops and have not developed complex variation patterns such as tea, coffee, rubber and other plantation crops.
Oligocentric crops	:	These crops are ancient ones with wide spread dispersal providing ample time and space for complexities to evolve and show nodes of high variability in several areas. Eg. Wheat
Noncentric crops	:	These are the crops derived from progenitors with wide dispersals and very often involved in multiple domestications like sorghum.
Diversity	:	The variety and variability present in a species or a cultivar or a land race.

VAVILOV'S THEORIES ON CENTER OF ORIGIN

A Russian botanist N.I. Vavilov was an extensive phytogeographer and global thinker and focussed an improving a human condition of the Russian federation. He followed Charles Darwin ideas and was influenced by his book on ``Variation of animals and plants under domestication'' and realized that the plant species were not fixed but they evolve over time. The Charles Darwin theory of evolution acted as a basis for vavilov Center of origin as indicated by several authors (Dorofeyev and Filatenko, 1992). Vavilov identified three major theories based on his readings and intuition through his field expeditions followed by plant evaluation.

i. Law of homologous variation

ii. Center of origin for cultivated crop plants

iii. Concept of genetic erosion

These theories have influenced greatly to the global plant science, breeding followed by conservation strategies.

SALIENT FEATURES OF VAVILOV'S EXPLORATIONS

The Russian botanist has evolved interest particularly in the science on ancient agricultural civilization on the mountainous region which he believed that these sites would be the major sites of plant diversity. Vavilov and his team have undertaken more than 100 plant expeditions during the period between 1920 and 1940. During these two decades of exploration, the team has brought more than 250000 accessions representing more than 10,000 species and assembled at the All Union Institution of Plant Industry. During the journey of exploration, Vavilov visited five continents and entered into almost all inaccessible mountains, deserts and forests. Through this journey, he realized the following points:

- The plant diversity was not distributed equally around the world but it was focussed only in specific region.
- He also realized that many such exploration need to be undertaken to obtain more genetic resources in order to validate the theoretical concept of Center of origin.

VAVILOV'S CENTER OF ORIGIN

Based on the explorations and expeditions made by Vavilov's and his team, he has proposed 5 Centers of origin in 1924 and later into 8 Centers in 1935. The major 8 Centers of origin and sub Centers of origin are furnished below:

S.NO	CENTER OF ORIGIN	COUNTRIES REPRESENTED
1.	**Chinese Center**	Central and Western China
2.	**Hindustan Center** Vavilo separated Hindustan center of origin into two sub centers **a. Indo Burma Center** **b. Indo-Malayan Center**	Assam and Burma, but not Northwest India, Punjab, nor Northwest Frontier Provinces. Indo-China and the Malay Archipelago.
3.	**Central Asiatic Center**	Punjab, Kashmir, Pakistan, Afghanistan and South-Western USSR.

S.NO	CENTER OF ORIGIN	COUNTRIES REPRESENTED
4.	**Near-Eastern Center**	Middle-east countries like Turkey, Iran, Israel and Plants
5.	**Mediterranean Center**	Countries around the Mediterranean sea
6.	**Abyssinian Center**	Embracing Ethiopia and Eritrea
7.	**Central American Center**	An isolated part in South Mexico and neighbouring countries
8.	**South American Center** Vavilov separated three sub Centers **a. Chiloe Center** **b. Brazilian-Paraguayan Center** **c. Peru Center**	Peru, Ecuador, Bolivia and Neighbouring islands

(*Source:* Singh.,1994)

These centers of origin were housed with wide range of crop plants like cereals, pulses, legumes, tubers, vegetables, fruits, nuts, sugar plants, drug, fibre plants and forest trees. The major plants which originated from these centers are furnished below :

1. CHINESE CENTER

a. Cereals and Legumes

Common Name	Scientific Name
Broomcorn millet	*Panicum miliaceum*
Italian millet	*Panicum italicum*
Kaoliang	*Andropogon sorghum*
Buckwheat	*Fagopyrum esculentum*
Hull-less barley	*Hordeum hexastichum*
Soybean	*Glycine max*
Adzuki bean	*Phaseolus angularis*
Velvet bean	*Stizolobium hassjoo*

b. Roots, Tubers, and Vegetables

Common Name	Scientific Name
Chinese yam	*Dioscorea batatas*
Radish	*Raphanus sativus*
Chinese cabbage	*Brassica chinensis, B. pekinensis*
Onion	*Allium chinense, A. fistulosum, A. pekinense*
Cucumber	*Cucumis sativus*

c. Fruits and Nuts

Common Name	Scientific Name
Pear	*Pyrus serotina*
Chinese apple	*Malus asiatica*
Peach	*Prunus persica*
Apricot	*Prunus armeniaca*
Cherry	*Prunus pseudocerasus*
Walnut	*Juglans sinensis*
Litchi	*Litchi chinensis*

d. Sugar, Drug, and Fiber Plants

Common Name	Scientific Name
Sugarcane	*Saccharum sinense*
Opium poppy	*Papaver somniferum*
Ginseng	*Panax ginseng*
Camphor	*Cinnamomum camphora*
Hemp	*Cannabis sativa*

2. HINDUSTAN CENTER

A. Indo-Burma center

a. Cereals and Legumes

Common Name	Scientific Name
Rice	*Oryza sativa*
Chickpea	*Cicer arietinum*
Pigeon pea	*Cajanus indicus*
Urd bean	*Phaseolus mungo*
Mung bean	*Phaseolus aureus*
Rice bean	*Phaseolus calcaratus*
Cowpea	*Vigna sinensis*

b. Vegetables and Tubers

Common Name	Scientific Name
Eggplant	*Solanum melongena*
Cucumber	*Cucumis sativus*
Radish	*Raphanus caudatus*
Taro	*Colocasia antiquorum*
Yam	*Dioscorea alata*

c. Fruits

Common Name	Scientific Name
Mango	*Mangifera indica*
Orange	*Citrus sinensis*
Tangerine	*Citrus nobilis*
Citron	*Citrus medica*
Tamarind	*Tamarindus indica*

d. Sugar, Oil, and Fiber Plants

Common Name	Scientific Name
Sugar cane	*Saccharum officinarum*
Coconut palm	*Cocos nucifera*
Sesame	*Sesamum indicum*
Safflower	*Carthamus tinctorius*
Tree cotton	*Gossypium arboreum*
Oriental cotton	*Gossypium nanking*
Jute	*Corchorus capsularis*
Crotalaria	*Crotalaria juncea*
Kenaf	*Hibiscus cannabinus*

e. Spices, Stimulants, Dyes, and Miscellaneous

Common Name	Scientific Name
Hemp	*Cannabis indica*
Black pepper	*Piper nigrum*
Gum Arabic	*Acacia arabica*
Sandalwood	*Santalum album*
Indigo	*Indigofera tinctoria*
Cinnamon tree	*Cinnamomum zeylanicum*
Croton	*Croton tiglium*
Bamboo	*Bambusa tulda*

B. Indo-Malayan Center

a. Cereals and Legumes

Common Name	Scientific Name
Job's tears	*Coix lacryma*
Velvet bean	*Mucuna utilis*

b. Fruits

Common Name	Scientific Name
Pummelo	*Citrus grandis*
Banana	*Musa cavendishii,* *M. paradisiaca,* *H. sapientum*
Breadfruit	*Artocarpus communis*
Mangosteen	*Garcinia mangostana*

c. Oil, Sugar, Spice, and Fiber Plants

Common Name	Scientific Name
Candlenut	*Aleurites moluccana*
Coconut palm	*Cocos nucifera*
Sugarcane	*Saccharum officinarum*
Clove	*Caryophyllus aromaticus*
Nutmeg	*Myristaca fragrans*
Black pepper	*Piper nigrum*
Manila hemp	*Musa textilis*

3. CENTRAL ASIATIC CENTER

a. Grains and Legumes

Common Name	Scientific Name
Common wheat	*Triticum vulgare*
Club wheat	*Triticum compactum*
Shot wheat	*Triticum sphaerocoecum*
Pea	*Pisum sativum*
Lentil	*Lens esculenta*
Horse bean	*Vicia faba*
Chickpea	*Cicer arientinum*
Mung bean	*Phaseolus aureus*
Mustard	*Brassica juncea*
Flax	*Linum usitatissimum*
Sesame	*Sesamum indicum*

b. Fiber Plants

Common Name	Scientific Name
Hemp	*Cannabis indica*
Cotton	*Gossypium herbaceum*

c. Vegetables

Common Name	Scientific Name
Onion	*Allium cepa*
Garlic	*Allium sativum*
Spinach	*Spinacia oleracea*
Carrot	*Daucus carota*

d. Fruits

Common Name	Scientific Name
Pistacia	*Pistacia vera*
Pear	*Pyrus communis*
Almond	*Amygdalus communis*
Grape	*Vitis vinifera*
Apple	*Malus pumila*

4. NEAR-EASTERN CENTER

a. Grains and Legumes

Common Name	Scientific Name
Einkorn wheat	*Triticum monococcum*
Durum wheat	*Triticum durum*
Poulard wheat	*Triticum turgidum*
Common wheat	*Triticum vulgare*
Oriental wheat	*Triticum orientale*
Persian wheat	*Triticum persicum*
Two-row barleys	*Hordeum distichum, H. nutans*
Rye	*Secale cereal*
Mediterranean oats	*Avena byzantina*
Common oats	*Avena sativa*
Lentil	*Lens esculenta*
Lupine	*Lupinus pilosus, L. albus*

b. Forage Plants

Common Name	Scientific Name
Alfalfa	*Medicago sativa*
Persian clover	*Trifolium resupinatum*
Fenugreek	*Trigonella foenum graecum*
Vetch	*Vicia sativa*
Hairy vetch	*Vicia villosa*

c. Fruits

Common Name	Scientific Name
Fig	*Ficus carica*
Pomegranate	*Punica granatum*
Apple	*Malus pumilo*
Pear	*Pyrus communis* and others
Quince	*Cydonia oblonga*
Cherry	*Prunus cerasus*
Hawthorn	*Crataegus azarolus*

5. MEDITERRANEAN CENTER

a. Cereals and Legumes

Common Name	Scientific Name
Durum wheat	*Triticum durum expansum*
Emmer	*Triticum dicoccum*
Polish wheat	*Triticum polonicum*
Spelt	*Triticum spelta*
Mediterranean oats	*Avena byzantina*
Sand oats	*Avena brevis*
Canarygrass	*Phalaris canariensi*
Grass pea	*Lathyrus sativus*
Pea	*Pisum sativum*
Lupine	*Lupinus albus*

b. Forage Plants

Common Name	Scientific Name
Egyptian clover	*Trifolium alexandrinum*
White Clover	*Trifolium repens*
Crimson clover	*Trifolium incarnatum*
Serradella	*Ornithopus sativus*

c. Oil and Fiber Plants

Common Name	Scientific Name
Flax	*Linum usitatissimum*, and wild *L. angustifolium*
Rape	*Brassica napus*
Black mustard	*Brassica nigra*
Olive	*Olea europaea*

d. Vegetables

Common Name	Scientific Name
Garden beet	*Beta vulgaris*
Cabbage	*Brassica oleracea*
Turnip	*Brassica campestris, B. napus*
Lettuce	*Lactuca sativa*
Asparagus	*Asparagus officinalis*
Celery	*Apium graveolens*
Chicory	*Cichorium intybus*
Parsnip	*Pastinaca sativa*
Rhubarb	*Rheum officinale*

e. Ethereal Oil and Spice Plants

Common Name	Scientific Name
Caraway	*Carum carvi*
Anise	*Pimpinella anisum*
Thyme	*Thymus vulgaris*
Peppermint	*Mentha piperita*
Sage	*Salvia officinalis*
Hop	*Humulus lupulus*

6. ABYSSINIAN CENTER

a. Grains and Legumes

Common Name	Scientific Name
Abyssinian hard wheat	*Triticum durum abyssinicum*
Poulard wheat	*Triticum turgidum abyssinicum*
Emmer	*Triticum dicoccum abyssinicum*
Polish wheat	*Triticum polonicum abyssinicum*
Barley	*Hordeum sativum*
Grain sorghum	*Andropogon sorghum*
Pearl millet	*Pennisetum spicatum*
African millet	*Eleusine coracana*
Cowpea	*Vigna sinensis*
Flax	*Linum usitatissimum*

b. Miscellaneous

Common Name	Scientific Name
Sesame	*Sesamum indicum*
Castor bean	*Ricinus communis*
Garden cress	*Lepidium sativum*
Coffee	*Coffea arabica*
Okra	*Hibiscus esculentus*
Myrrh	*Commiphora abyssinicia*
Indigo	*Indigofera argente*

7. CENTRAL AMERICAN CENTER

a. Grains and Legumes

Common Name	Scientific Name
Maize	*Zea mays*
Common bean	*Phaseolus vulgaris*
Lima bean	*Phaseolus lunatus*
Tepary bean	*Phaseolus acutifolius*
Jack bean	*Canavalia ensiformis*
Grain amaranth	*Amaranthus paniculatus*

b. Melon Plants

Common Name	Scientific Name
Malabar gourd	*Cucurbita ficifolia*
Winter pumpkin	*Cucurbita moshata*
Chayote	*Sechium edule*

c. Fiber Plants

Common Name	Scientific Name
Upland cotton	*Gossypium hirsutum*
Bourbon cotton	*Gossypium purpurascens*
Chayote	*Sechium edule*

d.

Common Name	Scientific Name
Sweetpotato	*Ipomea batatas*
Arrowroot	*Maranta arundinacea*
Pepper	*Capsicum annuum, C. frutescens*
Papaya	*Carica papaya*
Guava	*Psidium guayava*
Cashew	*Anacardium occidentale*
Wild black cherry	*Prunus serotina*
Cochenial	*Nopalea coccinellifera*
Cherry tomato	*Lycopersicum cerasiforme*
Cacao	*Theobroma cacao, Nicotiana rustica*

8. SOUTH AMERICAN CENTER

a. Root Tubers

Common Name	Scientific Name
Andean potato	*Solanum andigenum*
Edible nasturtium	*Tropaeolum tuberosum*

b. Grains and Legumes

Common Name	Scientific Name
Starchy maize	*Zea mays amylacea*
Lima bean	*Phaseolus lunatus*
Common bean	*Phaseolus vulgaris*

c. Root Tubers

Common Name	Scientific Name
Edible canna	*Canna edulis*
Potato	*Solanum phureja*

d. Vegetable Crops

Common Name	Scientific Name
Pepino	*Solanum muricatum*
Tomato	*Lycopersicum esculentum*
Ground cherry	*Physalis peruviana*
Pumpkin	*Cucurbita maxima*
Pepper	*Capsicum frutescens*

e. Fiber Plants

Common Name	Scientific Name
Egyptian cotton	*Gossypium barbadense*

f. Fruit and Miscellaneous

Common Name	Scientific Name
Passion flower	*Passiflora ligularis*
Guava	*Psidium guajava*
Heilborn	*Carica candamarcensis*
Quinine tree	*Cinchona calisaya*
Tobacco	*Nicotiana tabacum*

8 A. CHILOE CENTER

Common Name	Scientific Name
Common potato	*Solanum tubersum*
Wild strawberry	*Fragaria chiloensis*

8B. BRAZILIAN-PARAGUAYAN CENTER

Common Name	Scientific Name
Manioc	*Manihot utilissima*
Peanut	*Arachis hypogaea*
Rubber tree	*Hevea brasiliensis*
Pineapple	*Ananas comosa*
Brazil nut	*Bertholletia excelsa*
Cashew	*Anacardium occidentale*
Purple granadilla	*Passiflora edulis*

TYPES OF CENTERS OF DIVERSITY

The Centers of crop diversity are of three types *viz.,*

1. Primary Centers of diversity
2. Secondary Centers of diversity
3. Micro-Centers.

1. Primary Centers of Diversity

Primary Centers are regions of vast genetic diversity of crop plants. These are original homes of the crop plants which are generally uncultivated areas like, mountains, hills, river valleys, forests, etc. Main features of these Centers are given below.

i. They have wide genetic diversity.

ii. Have large number of dominant genes.

iii. The planets exhibit mostly wild characters.

iv. The crop plants exhibit less crossing over.

v. In the primary centre, predominantly natural selection operates.

2. Secondary Centers of Diversity

Vavilov suggested that valuable forms of crop plants are found far away from their primary area of origin, which he called secondary Centers of origin or diversity. These are generally the cultivated areas and have following main features.

i. Have lesser genetic diversity than primary centers.

ii. Have large number of recessive genes

iii. Mostly have desirable characters.

iv. Exhibit more crossing over

v. Both natural and artificial selections operate in the secondary centers of diversity.

3. Micro-centers

In some cases, small areas within the centers of diversity exhibit tremendous genetic diversity of some crop plants. These areas are referred to as micro-Centers. Micro-centers are important sources for collection of valuable plant forms and also for the study of evolution of cultivated species. The main features of micro-centers are given below.

i. They represent small areas within the Centers of diversity.

ii. Micro-centers exhibit tremendous genetic diversity.

iii. The rate of natural evolution in micro-centers is faster than larger areas.

iv. They are important sites for the study of crop evolution.

LIMITATIONS OF VAVILOV'S EXPLORATION

The Vavilov's concept of centers of origin on cultivated plants exhibited certain limitations and demanded adequate modifications. Some of his views which are contradicted are furnished below:

- The major view of Vavilov is that the region with higher genetic diversity of a species is also the centers of origin of that particular species. However this view of Vavilov's is debated because, several species observed in various region exhibited differences in centers of origin and genetic diversity.

  Example: Maize and Tomato.

- The second view of Vavilov is that most centers of origin of cultivated species is confined to mountains and small hills in tropical and sub tropical regions. However the recent evidences indicate that plains are also acting as a centers for many cultivated plants.

- Another major limitations is presence of more than one centers of origin for a crop which is different from the one suggested by the Vavilov.

SUMMARY

The centers of origin for many cultivated crops play a significant role as a resource base for domestic utility and deployment for ongoing breeding programme. The concept and theory established by Vavilov's have become guiding principles for plant breeding not only for domestication but also for global plant genetic resources conservation. His theory of homologous variation is now applied in the form of molecular genetics which is pronounced as a central dogma in plant breeding and evolution. His second theory of centers of origin is also now organized as centers of diversity and exhibited greater utility of deploying species for direct use by the human beings. Above all, his efforts on conservation and to control genetic erosion due to domestication of few cultivars have received greater attention across the world and insisted conservation strategies by controlling genetic vulnerability and genetic wipe out. In a holistic preview, this chapter has emphasized the scope, potential, opportunities and limitations of Vavilov's theories on centers of origin which may have profound utility in tree breeding and improvement programme in order to recognise the existing centers of origin for various tree crops and to identify new centers of origin for various tree crops and to identify new centers of origin for tree crops which is not currently established.

3

POLLINATION AND REPRODUCTION

INTRODUCTION

The pollination and its associated reproduction are very important in breeding and improvement of Forest Genetic Resources. The pollination is the transfer of pollen grains from the anther (male flower part) to the stigma of the plant (female flower part) which leads to fertilization. Pollination is the very important process which has been coordinated and carried out over ages as plants co-evolved with animals in the same habitats. This process of co-existence and co-evolution of plants and animals lead to the development of pollination syndrome. Pollination syndrome is nothing but the symbiotic relationship between the plants and the pollinating agents and both the organisms mutually benefit through this process. The pollination is very important for fertilization followed by successful seed production which ensures natural and artificial regeneration of several forest trees. This chapter elaborates different types of flowers, pollination along with the mechanisms to create basic understanding of the pollination process to the graduating scholars.

DEFINITION OF TERMINOLOGIES

Staminate flowers	:	Staminate flowers are flowers that only have stamens and are also known as androecious flowers.
Pistillate flowers	:	Pistillate flowers are flowers that only have pistils and are also known as unisexual flower.
Hermaphrodite flowers	:	A hermaphroditic flowers is one that produces both male and female reproduction systems.

Hermaphroditic flowers are largely self-pollinating and true bisexual plants.

Pollination : It is the process of transfer of pollen from the anther to the stigma of the flower leads to fertilization and results in zygotes formation

Reproduction : It is the process in which the living organism give rises to the offspring are similar kind.

or

It is the process of multiplication of living organisms.

Autogamy : Pollen grain falling on the stigma of the same flower or of a different flower but of the same plant is known as Autogamy. It is also known as self pollination.

Allogamy : The pollen grains from flowers of one plant are transmitted to the stigmas of flowers of another plant is known as Allogamy. It is also known as cross pollination.

Apomixis : Apomixis is the production of seeds without the fertilization of the ovule.

Pollination syndromes : Pollination syndromes are suites of flower traits that have evolved in response to natural selection imposed by different pollen vectors, which can be abiotic (wind and water) or biotic, such as birds, bees, flies, and so forth.

STRUCTURE OF FLOWER

The reproduction of angiosperm is carried out by the flowers which is formed due to a set of highly modified leaves with fertile a sterile

Characters

- The sterile leaves are formed by two whorls viz., calyx and corolla. The calyx is present is the outer layer which composed of sepals and the corolla is formed inside and is made up of petals. This sterile flower parts help to protect the fertile leaves which composed of male and female reproductive organism and are defined as androecium (male) and gynoecium (female) respectively.

- The sterile flower parts also play a vital role to create signals and attract the pollinators flower visitors through visual and recent related ques.
- The presence of one or two whorls depends on the species.
- The structures of a general flower along with parts are presented in the Fig. 3.1.

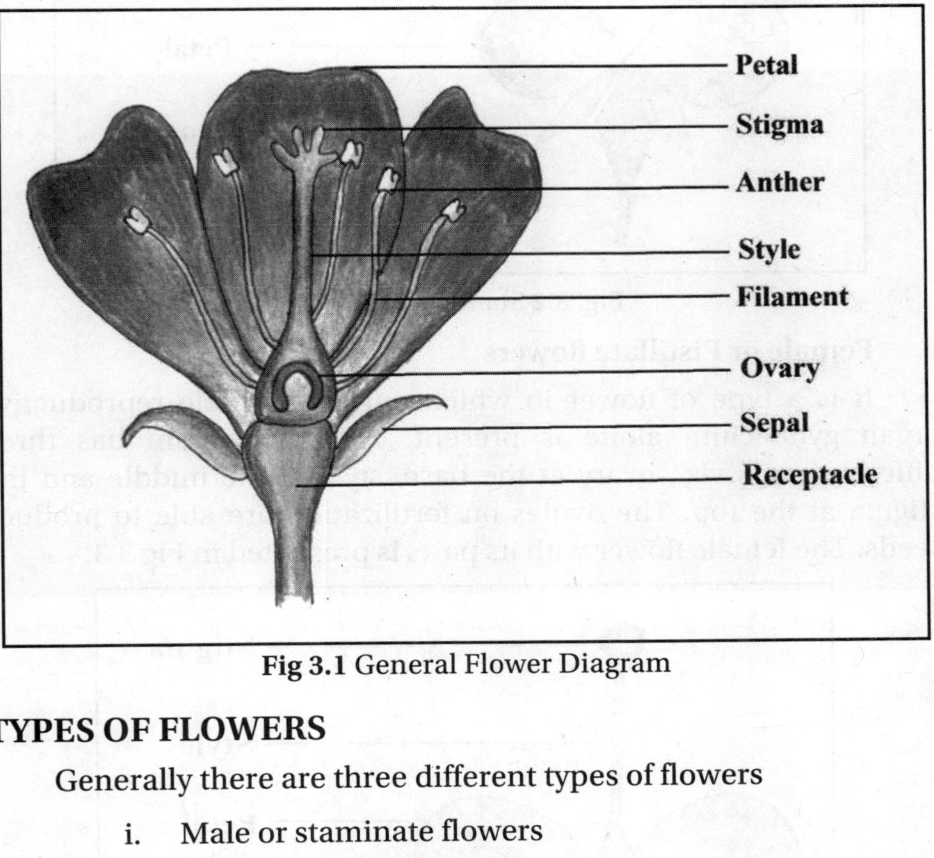

Fig 3.1 General Flower Diagram

TYPES OF FLOWERS

Generally there are three different types of flowers

i. Male or staminate flowers

ii. Female or Pistillate flowers

iii. Hermaphrodite flowers

i. Male or staminate flowers

This type of flowers contained only androecium which is formed by a group of stamens. Each stamen comprised a filament and the top of the filament anthers are present. Anthers are housed with wide range of pollen grains. The staminate flower is presented in Fig 3.2.

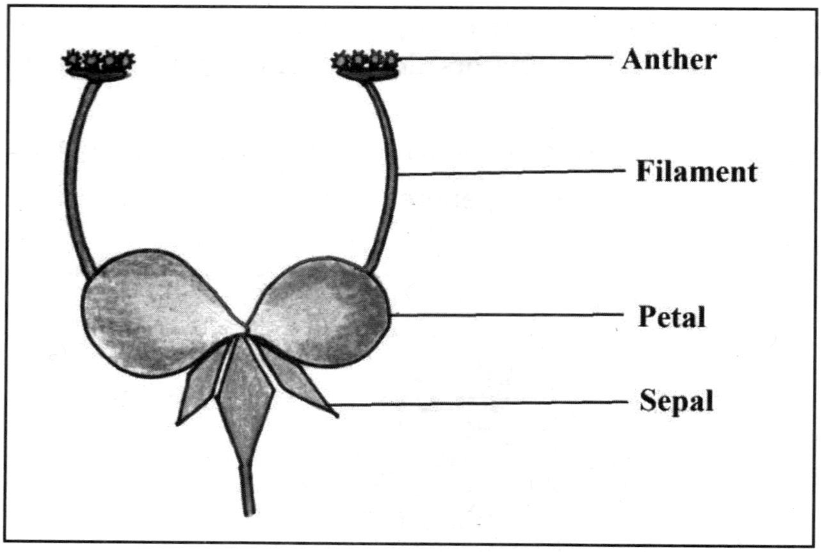

Fig. 3.2 Staminate Flower

ii. Female or Pistillate flowers

It is a type of flower in which only the female reproductive organ gynoecium alone is present. This gynoecium has three different parts viz., ovary at the base, style at the middle and the stigma at the top. The ovules on fertilization are able to produce seeds. The female flower with its parts is presented in Fig 3.3.

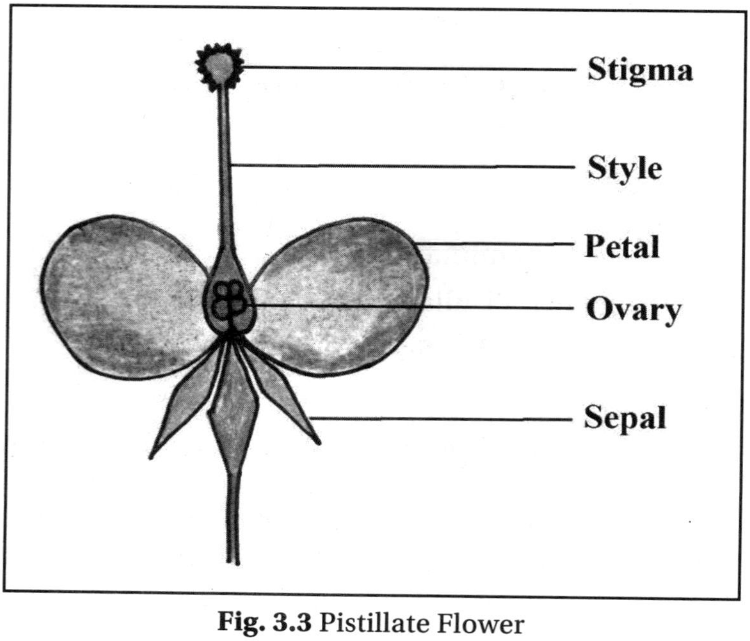

Fig. 3.3 Pistillate Flower

iii. Hermaphrodite Flower

In this type of flower both male and female reproductive organs namely the androecium and the gynoecium are present in the same flower and both of them are functional. The hermaphrodite flowers and their flower parts are furnished in Fig 3.4.

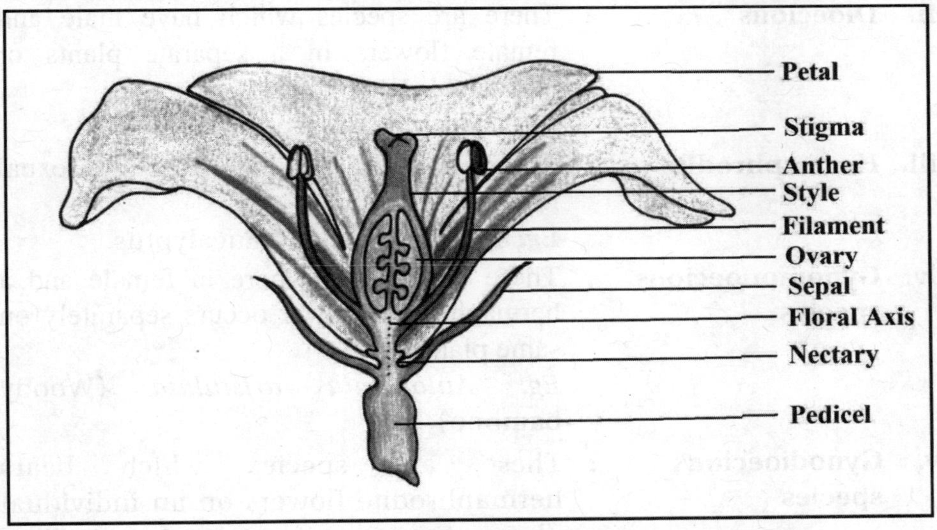

Fig. 3.4a. Hermaphrodite Flower

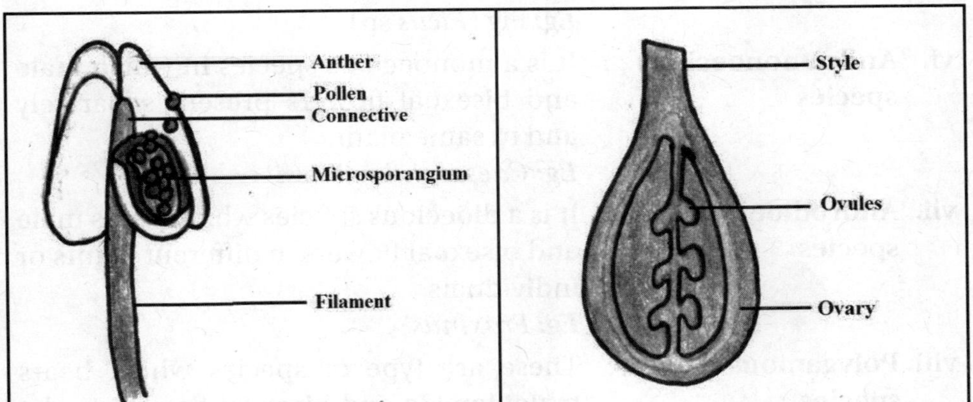

Fig. 3.4b. Stamen and Ovary of Hermaphrodite Flower

FLOWER TYPE DISTRIBUTION AMONG SPECIES

i. **Monoecious** : There are species which have male and female flowers in the same plant or individuals.
Eg: Jatropha

ii. **Dioecious** : There are species which have male and female flowers in a separate plants or individuals.
Eg: Casurina, Simarouba

iii. **Hermaphrodite** : These are species which posses bisexual flowers.
Eg: Sandal, Teak, and Eucalyptus.

iv. **Gynomonoecious species** : These are species where in female and a hermaphrodite flower occurs separately on same plants.
Eg: Aulonemia aristulata (Woody bamboo)

v. **Gynodioecious species** : These are species which bears hermaphrodite flowers on an individual plant that contains mostly pistillate flowers.
Eg: Fig (*Ficus* sp)

vi. **Andromonoecious species** : It is a monoecious species in which male and bisexual flowers present separately and in same plant.
Eg: Caesalpinia gilliesii

vii. **Androdioecious species** : It is a dioecious species which bears male and bisexual flowers in different plants or individuals.
Eg: Fraxinus

viii. **Polygamous species** : These are type of species which bears male, female and bisexual flowers in the same plant.
Eg: Garcinia spp.

SEXUAL EXPRESSION OF TROPICAL AND TEMPERATE TREES

The sexual expression varies with species particularly among the tropical and temperate species. In most vegetation complex, over

80% of the species are present with hermaphrodite (Bisexual) flowers and 10% of the species present with separated sexual expression and the remaining 10% exhibited wide variation with functional and non-functional reproductive organs. The sexual expression of tropical and temperate forest species are furnished below.

S.No	Sexual Expression	Frequency of Species	
		Tropical Forest	Temperate Forest
1	Hermaphrodite	68%	7%
2	Dioecious	10%	74%
3	Monocious	22%	19%

Source: Bawa (1974)

TYPES OF POLLINATION

There are three types of pollination

a. Self pollination
b. Cross pollination
c. Often cross pollination

a. SELF-POLLINATION

Pollen grains falling on the stigma of the same flower or of a different flower but of the same plant leads to self-pollination. These plants are called self pollinated crops or autogamous plants.

Eg. Oak

b. CROSS-POLLINATION

The pollen grains from flowers of one plant are transmitted to the stigmas of flowers of another plant is known as cross pollination or allogamy. These kinds of plants are called cross pollinated crops or allogamous plants.

Eg. Most tree species as indicated below

S.No	Species	Pollinators	Reference
1	*Amorphophalus* Species (Endemic Species of Western Ghat)	Beetles & Bees	Puneker & Kumeran, 2010
2.	*Cassia fistula*	Carpenter Bees	Murali , 1993

S.No	Species	Pollinators	Reference
3.	*Pittosporum dasycaulon* (Rare self incompatible medicine plant found in Western Ghat)	*Apis cerana* & *Apis dorsata*	Gopala Krishnan & Thomas, 2014
4.	*Myristica Dactloides* (Vulnerable swarm species found in Western Ghat)	Trips, Beetles, Bees and Flies	Sharma & Shivanna, 2011
5.	*Terminalia pallida* (endemic and medicine tree)	*Apis dorsata* and Wasps	Raju , 2012
6.	*Pavetta tomentosa* (Keystone shrub species found in Southern Eastern Ghat)	Butterflies and moths	Raju and Rao, 2016
7.	*Callistemon citrinus*	Animals and insects	Chauhan, 2017
8.	*Pterocarpus santalinus* (Indian Eastern Ghat)	*Apis dorsata*	Rao and Raju, 2002
9.	*Aquilaria* Spp.	20 different insect species	Sohertono and Newton, 2001
10.	*Shorea siamensis*	Trigona	Ghazoul and McLeash, 2001

c. Often Cross-Pollinated Species

The often cross pollination exceeds 5 per cent and may reach upto 30 per cent. Such species are generally known as often cross-pollinated species. The genetic architecture of such crops is intermediate between those of self-pollinated and cross-pollinated species. In often cross pollination species, breeding methods suitable for both of them may be profitably applied. But often hybrid varieties are superior to others.

I. SELF POLLINATION

Several tree species reproduce by self-pollination. Self-pollinated species are believed to have originated from cross-pollinated ancestors. These species as a rule, must have hermaphrodite flowers. But in most of these species, self-pollination is not complete and cross-pollination may occur up to 5%. The degree of cross-pollination in self-pollinated species is affected by several factors, e.g., variety, environmental conditions like temperature and humidity, and location.

Mechanisms Promoting Self-Pollination

There are various mechanisms that promote self-pollination, which are generally more efficient than those promoting cross-pollination. These mechanisms are briefly summarized below.

- ***Cleistogamy.*** In this case, flowers do not open at all. This ensures complete self-pollination since foreign pollen cannot reach the stigma of a closed flower.
- ***Chasmogamy.*** In some species, the flowers open, but only after pollination has taken place.
- Pollination generally occurs after the flowers open. But the position of anthers in relation to stigmas ensures self-pollination.
- In some species, flowers open but the stamens and the sigma are hidden by other floral organs.
- In a few species, stigmas become receptive and elongate through the stamina columns. This ensures predominant self-pollination.

Genetic Consequences of Self-Pollination

- Self-pollination leads to a very rapid increase in homozygosity.
- Populations of self-pollinated species are highly homozygous.

II. CROSS-POLLINATION

In cross-pollinating species, the transfer of pollen from a flower to the stigmas of the others may be brought about by wind (*anemophily),* water (*hydrophily)* or insects (*entomophily*). In many species, a small amount (up to 5-10 per cent) of selfing may also occur.

Mechanism Promoting Cross Pollination

There are several mechanisms that facilitate cross pollination; these mechanisms are described briefly.

i. **Dicliny** or **unisexuality:** It is a condition, in which the flowers are either staminate (male) or pistillate (female).

a. **Monoecy:** Staminate and pistillate flowers occur in the same plant either in the same inflorescence.

b. **Dioecy:** The male and female flowers are present on different plants.

   i.e., the plants in such species are either male or female.

   E.g., Casuarina and Simarouba

   - In general, the sex is governed by a single gene.
   - In some cases, there are hermaphrodite plants in addition to male and female plants, and a number of intermediate forms may also occur.

ii. Dichogamy: Stamens and pistils of hermaphrodite flowers may mature at different times, thereby, facilitating cross-pollination.

a. **Protogyny:** Pistils mature before stamens.

b. **Protandry:** Stamens mature before pistils.

Sometime, stigmas are covered with a waxy film. The stigma does not become receptive until this waxy film is broken. The waxy membrane is broken by the visit of honey bees, which also effect cross-pollination.

A combination of two or more of the above mechanisms may occur in some species. This improves the efficiency of the system in promoting cross-pollination.

iii. Self-Incompatibility: It refers to the failure of pollen from a flower to fertilize the same flower or other flowers on the same plant. Self-incompatibility is of two types: Sporophytic and gametophytic. In both the cases, flowers do not set seed on selfing. It is highly effective in preventing self-pollination.

iv. Male sterility: Male sterility refers to the absence of functional pollen grains in otherwise hermaphrodite flowers. Male sterility is not common in natural populations. But it is of great value in experimental populations, genetic and cytoplasmic. Cytoplasmic male sterility is termed cytoplasmic genetic when restorer genes are known. In view of the importance of self-incompatibility and male sterility.

Genetic Consequences of Cross-Pollination

- Cross Pollination preserves and promotes heterozygosity in a population.
- Cross-pollinated species are highly heterozygous and show mild to severe inbreeding depression.
- Cross pollinated species exhibit a considerable amount of heterosis.

POLLINATORS

These are the agents of both biotic and abiotic factors responsible for transfer of pollen from the flower to stigma of the flower of the same or different plants.

TYPES OF POLLINATORS

i. Abiotic pollinators

They are otherwise called non living pollinators which include wind, water, gravity etc., which facilitates transfer of pollen to the stigma of the flower. They are further classified into the following types:

a. Anemophily	:	It is the type of abiotic pollination facilitated by wind. *Eg.* Poplar
b. Hydrophily	:	It is the type of abiotic pollination carried out by water. *Eg.* Terminalias, litoral and swamp species
c. Gravity	:	It is the abiotic pollination facilitated by gravity. *Eg.* Gyrocarpus

ii. Biotic pollinators

They are otherwise called as living pollinators wherein the transfer of pollen from male reproductive organ to the stigma of a female reproductive organ is facilitated by animals, insects and birds. Biotic pollinators are further classified into various types based on the actual biological organism involved in pollination process.

a. Entomophily	:	It is a type of biotic pollination carried out by insects *Eg.* Jatropha, Teak, Sandal, etc.

b. Cantharophily	:	It is a type of biotic pollination carried out by beetles ***Eg.*** *Myristica insipid, Diospyros pentamera*
c. Cheiropeterophily	:	it is a type of biotic pollination carried out by bats. ***Eg.*** Pollination in Agave, Madhuca and fig
d. Melittophily	:	It is a type of biotic pollination carried out by bees and it is otherwise termed as himopterophily ***Eg.*** Sandal and Jatropha
e. Myrmecophily	:	It is a type of biotic pollination carried out by ants ***Eg.*** Most tree crops
f. Necrocoleopterophily	:	It is a type of biotic pollination carried out by carrion beetle. ***Eg.*** Pollination in carrion flower
g. Phalaenophily	:	It is a type of biotic pollination carried out by moth. ***Eg.*** Pollination in *Lantana camara,* Plants from *Verbenaceae, Asclepiadaceae, Caesalpinacace, Apocynaceae, Rubiaceae, Solanaceae, Campanulaceae-Lobeloideae, Liliaceae, Caricaceae, Combretaceae, Nyctaginaceae, Pittosporaceae, Agavaceae, Convolvulaceae,* etc,.
h. Psychophily	:	It is a type of biotic pollination carried out by Butterflies ***Eg.*** *Stachytarpheta jamaicensis,* Plants from *Verbenaceae, Caesalpinaceae, Zingiberaceae, Liliacaceae, Boraginaceae, Nyctaginaceae, Orchidaceae, Rubiacaeae.*
i. Sapromyiophily	:	It is a type of biotic pollination carried out by Flies such as carrion, dung and mushroom fly ***Eg.*** Pollination in carrion flower

j. **Ornithophily**	:	It is a type of biotic pollination facilitated by birds *Eg.* Eucalyptus, Simarouba, Neem

FACTORS FACILITATING ANIMAL POLLINATION

Pollination and the reproduction of most tropical forest trees largely depend on the interaction between plants and animals which transfer pollen grains and facilitates the pollination process.

- Flower shape, colour and scent are the important factors which reveals significant information about the plant pollinator relationship and the composition of pollination guilds.
- The flower morphology particularly the size and shape play a vital role either to attract or repell the pollinators.
- The morphological and physiological factors also act as an indicator for attraction and repell by different agents
- The color prescription of the several species will act as a major factor for the pollination process.
- Some of the flowers produces aromatic compounds which are attracted by the insects and facilitates pollination.
- The feeding behaviors of some of the animals also play a vital role in facilitating pollination process.
- The prevailing wind in certain forest types and habitats act as a major source for pollination
- Some of the birds are attracted by nectors and colour which results in completion of pollination.
- The rotten smell like fruits facilitates pollination by bats.
- In coastal and swampy areas, water play a vital role in facilitating pollination process.

MODE OF REPRODUCTION

In general most of the tree species are reproduced by two means

1. Sexual Reproduction
2. Asexual Reproduction

1. Sexual Reproduction

It is the method of reproduction in which fusions of male and female gametes occurs which leads to formation of embryos. This sexual reproduction involves meiotic cell division, mitotic cell division, fertilization and zygote formation.

Process of Sexual Reproduction

- Firstly, the formation of microspores in anther and megaspores in ovules is done through meiotic cell division. This process of formation of spores is termed as sporogenesis.
- Secondly, the sperm cells are formed in microspores and egg cells are formed in megaspores through meiotic cell division. This process of formation of sperms and eggs is termed as gametogenesis.
- After this sporogenesis and gametogenesis the union of male and female gametes occurs through fertilization which leads to formation of zygotes.
- This zygotes is further developed into embryos.

2. Asexual Reproduction

It is termed as clonal or vegetative reproduction and does not involve meiosis and fertilization as happened in sexual reproduction. There are 2 types of asexual reproduction.

i. Vegetative Reproduction

ii. Apomixis

i. Vegetative Reproduction

The various methods of asexual vegetative reproduction are provided below.

S.No	Methods	Tree species
1.	Rooting of cuttings	Casuarina, Eucalyptus, Melia, Kadam
2.	Grafting	Pungam, Simarouba
3.	Layering	Most fruit trees
4.	Suckers and Rhizomes	Bamboos and Canes
5.	Root suckers	Melia, Sissoo
6.	Tubers	Wild medicinal plants

ii. Apomixis

Apomixis is the production of seeds without the fertilization of the ovule.

Mechanism and types of apomixes

In the apomixitic plant, the embryo is derived from mitotic division of megaspore mother cell or a somatic cell of the ovules. This apomixis is developed through various mechanisms based upon the origin and development of cell from which the embryo develops. The various forms of apomixis are furnished below

Parthenogenesis	It is the mechanism of development of embryo from the egg shell without undergoing fertilization (or) development of embryo directly from an unfertilized egg. The resulting plant is haploid.
Apogamy	It is the mechanism in which the embryo develops from haploid nuclei other than the egg (or) it is the process of formation of embryo from synergid or antipodal cells of embryo sac. The resulting plant is haploid.
Apospory	It is the mechanism in which the embryo develops from an unreduced egg in an embryo sac from a nuclear cell. The resulting plant is diploid.
Diplospory	It is the mechanism wherein the embryo develops from mitotic division of megaspore mother cell. The resulting plant is diploid.
Adventitious embryony	It is the mechanism wherein the embryo develops directly from a somatic cell of the nucellus. The resulting plant is diploid.

SUMMARY

Pollination and reproduction are very important and essential processes in multiplication and development of wide range of plant species. The forests, particularly the tropical forest which acts as a treasure house of plant diversity and is attributable due to inherent pollination mechanism. The observation on the process of pollination followed by reproduction and dispersal are very essential for foresters and tree breeders to understand the spatial distribution of tree species in an established vegetal formation. This will also help to augment the diversity and to protect the species which are threatened with vulnerability and endanger. The interaction between the plants and animals in the process of pollination syndrome will help to balance the entire ecosystem through the symbiotic and mutual associations. This pollination exhibited wide variation due to type and nature of flower, pollinating agents (both biotic and abiotic) both in the tropical and temperate forests. The pollination mechanism and the associated reproductive biology have been very well established for most temperate species. Such intensive reproductive biology studies in tropical tree species are dismally modest except few species. Hence this chapter on pollination and reproduction has been included to provide the basic understanding for those who are involved in tree breeding and improvement process.

4

METHODS OF TREE BREEDING

INTRODUCTION

Plant breeding has made significant achievement deploying different methods of breeding. However all the methods employed in crop breeding may not apply equally taking into consideration the large phenotypic nature of the trees. The tree breeders have several opportunities to deploy methods followed in crop breeding simultaneously or may combine several methods. The methods of tree breeding needs to be evaluated based on genetic and breeding principles coupled with the species and the traits of interest. In most tree breeding, it is commonly found that a breeding method alone may not be effective to achieve the objectives but may need combination of silvicultural practices. Above all, the most breeding programmes are all long time nature and demand huge funding support. Hence it is very important to choose a suitable breeding method which is efficient and simple taking into consideration the time and cost involved in tree breeding programmes. Under such circumstances unlike agricultural and other crop plants, tree breeding demands only few practically applicable methods. Among several breeding methods, selection in most tree species followed by hybridization in certain commercially important tree species are widely followed to improve the productivity and quality. This chapter elaborates major breeding methods deployed in tree breeding programme.

COMMON TREE BREEDING METHODS

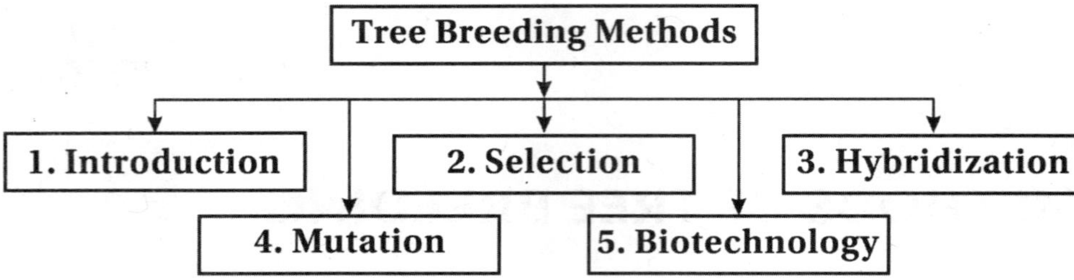

GENETIC RESOURCES FOR TREE BREEDING

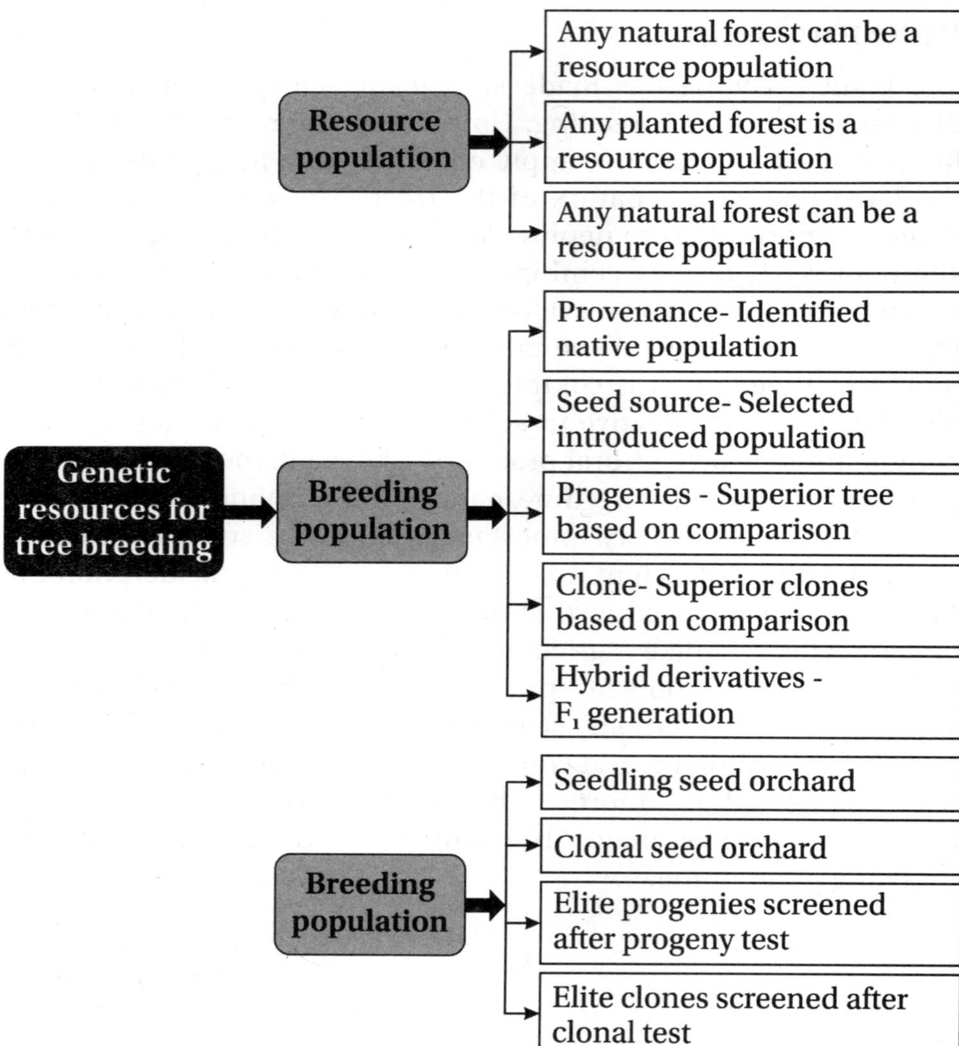

TREE BREEDING METHODS

1. INTRODUCTION OR IMPORTATION OF NEW SPECIES

Introduction of new species of various tree crops from different regions or different countries of the world has been an important method of tree breeding for several tree species. The introduction of new species is done to satisfy the demands of both domestic and industrial sector on a commercial basis. This besides, wide range of species has been introduced for ornamental purposes. Sometimes the new species are introduced for incorporation in hybridization programme with an objective of combining one or more desirable traits with local species.

The species which has been introduced into India for commercial and other purposes are furnished below.

Tree Species Introduced into India for commercial and other purposes

S.No	Common name	Scientific name	Origin	Purpose
1.	Mysore gum	*E. tereticornis*	Australia	Pulpwood
2.	Rose gum	*E. grandis*		Firewood, small timber, poles
3.	Grey iron bark	*E.paniculata*		
4.	Spotted gum	*E.maculata*		
5.	Lemon scented gum	*E. citriodora*		
6.	Blue gum	*E. globulus*		Pulp wood and fuel wood
7.	River Red gum	*E. camaldulensis*		Pulpwood
8.	Poplar/ Eastern cotton wood	*Populus deltoides*	North America	Multipurpose tree

S.No	Common name	Scientific name	Origin	Purpose
9.	Beef wood/she oaks	*Casuarina equisetifolia*	Australia and coastal south east Asia	Firewood
10.	Beef wood	*Casuarina junghuhniana*	Indonesia	Pulp and paper
11.	Subabul	*Leucaena leucocephala*	Hawaii, USA	Multipurpose tree/ fuelwood
12.	Australian wattle	*Acacia auriculiformis*	North of Australia	Plantation
13.	Israeli babool	*Acacia tortilis*	Israel	Afforestation programme in arid region
14.	Mesquite	*Prosopis juliflora*	Mexico, South America and Carribean	Afforestation
15.	Black wattle	*Acacia dealbata*	Australia	Firewood
16.	Silver oak	*Grevillea robusta*	Australia	Shade tree
17.	Radiata Pine	*Pinus radiata*	Central Coast of California and Mexico	Pulp and lumber
18.	Brown salwood	*Acacia mangium*	North eastern Australia, Indonesia and Papua guinea	Reforestation purpose
19.	Physic nut	*Jatropha curcas*	Central America and Mexico	Multipurpose shrub
20.	Gulmohar	*Delonix regia*	Madagascar.	Ornamental tree
21.	Paradise tree	*Simarouba glauca*	Southern florida, West Indies and Brazil	Plantation

S.No	Common name	Scientific name	Origin	Purpose
22.	Annatto	*Bixa orellana*	America	Medicinal pursposes
23.	Black wattle	*Acacia mearnsil*	Australia	plantation
24.	Indonesian Albizzia / Moluccan Albizzia	*Albizia falcataria*	Indonesia	Shade tree
25.	Paper mulberry	*Broussonetia papyrifera*	China and Japan	Paper printing
26.	Japanese cedar	*Crpytomeria japonica*	Japan	Mixture plantation
27.	Balsa wood	*Ochrama pyramidalis*	America	Plantation
28.	Rubber tree	*Hevea brasiliensis*	Brazil	Rubber
29.	Darwin silky oak	*Grevillea pteridifolia*	Australia	Reclaim bauxitemined areas
30.	Jerusalem thorn	*Parkinsonia aculeate*	Mexico	Afforestation of dry areas
31.	Caribean pine	*Pinus caribaea*	Br. Honduras	Plantation
32.	Patula pine	*Pinus patula*	Mexico	Plantation
33.	False acacia	*Robinia pseudoacacia*	North America	Reforest barren hills
34.	Rain tree	*Samanea saman*	Central America	Road side avenue
35.	White willow	*Salix alba*	Europe	Cricket bats
36.	Broad leaved sweiteinia	*Swietenia macrophylla*	C. America	Ornamental

S.No	Common name	Scientific name	Origin	Purpose
37.	Small leaved Mahogany	*Swietenia mahagani*	Cuba, Jamaica	Ornamental Avenue Tree
38.	Jacaranda	*Jacaranda mimosifolia*	Brazil and Argentina	Ornamental tree
39.	Indian Lavender	*Bursera peniciata*	Mexico	Plantation
40.	Cashew nut	*Anacardium occidentale*	Tropical America	Nut
41.	Australian black wood	*Acacia melanoxylon*	Tasmania, South Australia and Queensland	Avenue tree

Advantages of introduction of new species

- It is a very easy and quick method of developing new varieties especially when introduced material is used directly as a variety or after selection as that of eucalyptus and casuarina.
- Introduction is a good method of collection and conservation of germplasm of different tree species to protect the same from genetic erosion.
- This is an effective method of conserving those tree species which have been threatened by the danger of extinction. Such species can be saved by shifting them to other areas. Very classical example is Indian sandal which has been pronounced as a endangered species but found wider cultivation in Australia. It is now called as Australian seed source and Indian Provenance and thus act as an *ex-situ* conservation park besides catering the needs of breeders and industries demanding sandal wood.
- Sometimes the introduced species will be a fast grower compared to the local species and hence the introduced species may have greater adoptability.

Ex. Eucalyptus – largest plantation area in India

Casuarina – Dominant introduced species in coastal districts of India

Poplars – Largest introduced species in Indo-gangetic plains

- Introduced tree species exhibit wide variety of wood and wood products compared to the local species and hence more amenable for domestic and industrial utility.

  Ex. Poplars – Source of Plywood

  Eucalyptus – Source of Pulpwood and plywood

  Silver oak – Source of timber

- Some of the introduces tree species may act as a excellent source of breeding population to improve the local varieties

  Ex. *Populus deltoides* (Introduced) is used in improving the local poplar varieties like *Populus ciliata*

- Introduced species exhibit wider adoptability and hence more advantages to convert the barren land into a forest plantation.

  Ex. Eucalyptus, Subabul, wattles

Disadvantages of tree introductions

- The introduction of species and its establishment is a costly affair and time consuming process to test all the promising tree species on a wide range of edapho climatic conditions.
- Sometimes the fast growing nature of tree species may affect the growth and development of local species.
- Some of the introduced species like wattles and prosopis have been found as an invasive species due to their abundant seed production which lead to degradation of growth and development of local species. Sometimes the introduced species may alter the biodiversity and the local habitat which will affect both flora and fauna.
- The introduced species may sometime become a major threat to the biodiversity of an introduced habitat.
- Some of the introduced species may bring in several pest and diseases which may affect the growth and development of local species and landraces.

- There are also chances of failure of introduced species due to bioclimate variability thus resulting in failure of the entire tree breeding programme.
- Due to introduction of species leading to wide range of product development may affect the market potential of local species. Very classical example is introduced African mahogany and teak affect the market potential of Indian teak.

2. SELECTION

Selection is one of the most important breeding methods help to obtain significant amount of genetic gain quickly and inexpensively besides maintaining a broad genetic base. It is one of the most commonest methods in tree breeding and improvement of several tree species. The fundamental basis of selection is to exploit the existing natural variability. Since most tree crop are highly cross pollinated and the resultant base populations are heterogenous and heterozygous exhibit wider variability. Under such circumstances, selection is the only easy and best method to exploit the variability and to deploy them through various breeding programme to validate the variability which may be due to genetic in nature or may be present due to the influence of the environment.

The most common practices of selecting the plus tree is furnished below

Plus tree Selection

- **1. Candidate tree** → A tree that has been selected among candidate trees based on subjective grading due to desirable and superior phenotypic qualities and incorporated in further genetic test.
- **2. Superior / Plus tree** → A tree that has been recommended for production or breeding following grading. It has a superior phenotype for growth form wood quality and desired characters.
- **3. Elite Tree** → A term reserved for selected trees that have proven to be genetically superior by means of progeny testing. An elite tree is the winner from a selection programme.
- **4. Comparison tree** → Trees that are located in the same stand, which are of the same age growing on the same or better site as the selected tree and against which the selected tree is graded.

Population and selection method

S.No	Population	Method	Salient Features
1.	Even aged	Comparison tree method	• The stands are truly even-aged. • Candidate and comparison trees are available. • Selection of plus trees based on individual superiority in terms of clean bole, height, GBH, tree form, free from pest and diseases. • The selected plus tree should be superior to the neighbourer or comparison or check trees.

S.No	Population	Method	Salient Features
			• Selecting one tree for every thousand population at an escapement of 3m × 3m is recommended. However higher selection intensity is preferred to get best individual. • Two stage of selection programme is employed. • First stage is preliminary reporting of outstanding tree called as candidate trees. • Final phenotypic appraisal and approval as plus trees.
2.	Uneven Even aged	Regression method	• The stand is truly uneven aged with different ages. • The candidate trees is distributed sporadically • Comparison tree may not exists • Presence of mixed undesirable species • Candidate trees are selected based on phenotypic appraisal in terms of tree form, growth attributes, apical dominance, free from pest and diseases. • The selected trees are evaluated through Regression analysis. • The trait is plotted (Predominantly volume or any trait of interest) on the regression graph using proper age and site. • The trees which fall above the regression line are selected and the trees which fall below the Regression line are rejected.

S.No	Population	Method	Salient Features
			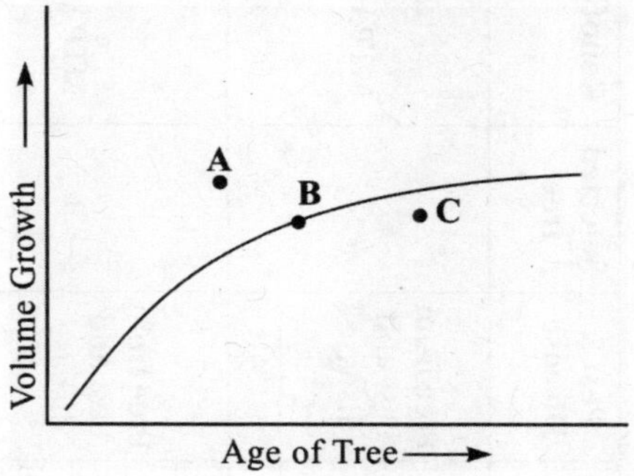

PRACTICAL EXAMPLE OF PLUS TREE SELECTION IN KADAM PLANTATIONS

A six year old Kadam plantation established using 20 different seed sources at an espacement of 3 m × 3 m at Forest College and Research Institute has been evaluated for selection of plus trees following the method of comparison or check tree method. The candidate trees are selected and compared with check trees and finally selected as a superior or plus tree. The various plus trees selected by using comparison tree method are furnished below.

Table 1. Biometric attributes of plus trees in *Neolamarckia cadamba*

Sampling	Comparison tree	Basal Girth (cm)	Girth @BH (cm)	Height (m)	Bole Straightness	Bark Texture	Branching Pattern	Pest & Disease	Selected tree	Genotype
PS 1	1	70	63.5	8	3	2	3	Free from Pest and disease	3	MTPAC 1
	2	43.5	32.5	6	5	1	2			
	3	86	65	8	5	2	2			
	4	56	42.5	8	4	2	1			
Mean		**63.87**	**50.87**	**7.5**						
PS 2	1	37	22.5	5.5	4	2	2	Free from Pest and disease	3	MTPAC 2
	2	43.5	32.5	6	5	1	2			
	3	66	48	9	5	2	2			
	4	38.5	30	6	5	3	2			
	5	54	41	7	4	2	2			
Mean		**47.8**	**34.8**	**6.7**						
PS 3	1	67	48.5	8.5	5	2	3	Free from Pest and disease	1	MTPAC 3
	2	64.5	48	8	4	3	3			
	3	45	39.5	7	4	2	3			
	4	41.5	28.5	5.5	4	3	2			
	5	50	39.5	7.5	4	2	3			
	6	40.5	30.5	6.5	5	1	3			
Mean		**51.41**	**39.08**	**7.16**						
PS 4	1	58	41	8	5	4	2	Free from Pest and disease	1	MTPAC 4
	2	44.5	33	6.5	5	2	3			
	3	32.5	29.5	6	5	2	1			
	4	37	26.5	5	4	3	3			
Mean		**43**	**32.5**	**6.37**						

Sampling	Comparison tree	Basal Girth (cm)	Girth @BH (cm)	Height (m)	Bole Straightness	Bark Texture	Branching Pattern	Pest & Disease	Selected tree	Genotype
PS 5	1	69.5	56	9	4	4	3	Free from Pest and disease	1	MTPAC 5
	2	44	34.5	7.5	4	2	2			
	3	38	28.5	4.5	4	3	2			
	4	52	41	8	3	2	3			
Mean		**50.87**	**40**	**7.25**						
PS 6	1	66.5	52	8.5	4	2	3	Free from Pest and disease	1	MTPAC 6
	2	44.5	30.5	7	4	1	2			
Mean		**55.5**	**41.25**	**7.75**						
PS 7	1	53	42.5	8.5	5	2	2	Free from Pest and disease	1	MTPAC 7
	2	47.5	35	7	3	3	2			
	3	47	36.5	7.5	4	2	1			
	4	47	33	8	3	2	1			
Mean		**48.62**	**36.75**	**7.75**						
PS 8	1	72.5	56	9.5	5	3	2	Free from Pest and disease	1	MTPAC 8
	2	74	53.5	8	3	3	3			
	3	58.5	50	8	3	4	3			
Mean		**68.33**	**53.16**	**8.5**						
PS 9	1	52.5	40	8.5	4	3	3	Free from Pest and disease	1	MTPAC 9
	2	40	32	8	3	2	1			
Mean		**46.25**	**36**	**8.25**						
PS 10	1	66.5	52	9	5	3	3	Free from Pest and disease	1	MTPAC 10
	2	34	26	6.5	3	2	1			
	3	33	26.5	6	4	2	2			
	4	38	28.5	7	4	3	2			
Mean		**42.87**	**33.25**	**7.12**						

Sampling	Comparison tree	Basal Girth (cm)	Girth @BH (cm)	Height (m)	Bole Straightness	Bark Texture	Branching Pattern	Pest & Disease	Selected tree	Genotype
PS 11	1	43	34.5	9.5	5	1	2	Free from Pest and disease	1	MTPAC 11
	2	27	19	5	5	1	1			
	3	32.5	21.5	6	4	2	2			
	4	36	29.5	7	4	2	1			
Mean		**34.62**	**26.12**	**6.87**						
PS 12	1	46	35.5	8	5	2	2	Free from Pest and disease	1	MTPAC 12
	2	40.5	30	7	4	1	2			
	3	44.5	34	6.5	3	2	1			
	4	36	25	6	4	2	3			
Mean		**41.75**	**31.12**	**6.87**						
PS 13	1	42	36.5	8	5	2	2	Free from Pest and disease	1	MTPAC 13
	2	29	19.5	5.5	4	3	1			
	3	36.5	25.5	5	3	2	2			
	4	33	26	7.5	4	1	2			
Mean		**35.12**	**26.87**	**6.5**						
PS 14	1	44.5	36	8.5	5	2	2	Free from Pest and disease	1	MTPAC 14
	2	37	25.5	5.5	4	2	2			
	3	44	29.5	6	4	1	3			
	4	39.5	29	7.5	4	2	1			
Mean		**41.25**	**30**	**6.87**						
PS 15	1	48.5	37.5	9	5	2	1	Free from Pest and disease	1	MTPAC 15
	2	45	36.5	8	4	2	1			
	3	40.5	33.5	7	4	2	1			
Mean		**44.66**	**35.83**	**8**						

Sampling	Comparison tree	Basal Girth (cm)	Girth @BH (cm)	Height (m)	Bole Straightness	Bark Texture	Branching Pattern	Pest & Disease	Selected tree	Genotype
PS 16	1	64.5	37.5	10	5	2	1	Free from Pest and disease	1	MTPAC 16
	2	41	30	6	4	2	2			
	3	52	37	8	4	2	3			
Mean		**52.5**	**34.83**	**8**						
PS 17	1	53	39	8.5	5	2	3	Free from Pest and disease	1	MTPAC 17
	2	42	30.5	7	4	2	1			
	3	38	28	7.5	4	2	1			
	4	32	23.5	6	4	2	1			
	5	38.5	26.5	8	5	2	1			
Mean		**40.7**	**29.5**	**7.4**						
PS 18	1	45.5	37	9	5	2	1	Free from Pest and disease	1	MTPAC 18
	2	47.5	31	8	4	2	3			
	3	44.5	30	7.5	4	2	3			
	4	45	33.5	6.5	4	2	3			
Mean		**45.62**	**32.87**	**7.75**						
PS 19	1	61	43	8.5	5	2	1	Free from Pest and disease	1	MTPAC 19
	2	34.5	27	7	4	2	1			
	3	34.5	24	6	5	2	2			
	4	37	28.5	6	5	2	1			
	5	48	34.5	8	4	1	3			
Mean		**43**	**31.4**	**7.1**						
PS 20	1	44.5	33.5	8	5	2	1	Free from Pest and disease	1	MTPAC 20
	2	40	27	6	4	3	2			
	3	33	24	6	3	2	1			
	4	41	28	7.5	4	2	3			
Mean		**39.62**	**28.12**	**6.87**						

Sampling	Comparison tree	Basal Girth (cm)	Girth @BH (cm)	Height (m)	Bole Straightness	Bark Texture	Branching Pattern	Pest & Disease	Selected tree	Genotype
PS 21	1	53.5	40.5	9	5	2	1	Free from Pest and disease	1	MTPAC 21
	2	33	24	6	3	2	1			
	3	49	33.5	7.5	4	2	3			
	4	41	28	7.5	4	2	3			
	5	46	34.5	8.5	4	2	3			
Mean		**44.5**	**32.1**	**7.7**						
PS 22	1	73.5	48	10	5	2	3	Free from Pest and disease	1	MTPAC 22
	2	53	38	8.5	4	2	3			
	3	31.5	21	6	4	2	2			
	4	55.5	30.5	7	3	2	3			
	5	47	21.5	6.5	4	2	3			
Mean		**52.1**	**31.8**	**7.6**						
PS 23	1	48.5	33	6.5	3	2	3	Free from Pest and disease	2	MTPAC 23
	2	81.5	51.5	9.5	5	2	3			
	3	39	28	7	4	2	3			
	4	40.5	28	7	4	2	3			
	5	30.5	23.5	6.5	4	2	3			
Mean		**48**	**32.8**	**7.3**						
PS 24	1	35	23.5	6	3	2	3	Free from Pest and disease	3	MTPAC 24
	2	46	35	8.5	4	2	3			
	3	48.5	38.5	8.5	5	2	1			
	4	52.5	37	7	4	2	3			
	5	50.5	32	8	4	2	3			
Mean		**46.5**	**33.2**	**7.6**						

Sampling	Comparison tree	Basal Girth (cm)	Girth @BH (cm)	Height (m)	Bole Straightness	Bark Texture	Branching Pattern	Pest & Disease	Selected tree	Genotype
PS 25	1	50.5	37	8	3	2	3	Free from Pest and disease	4	MTPAC 25
	2	49.5	32	7	4	2	3			
	3	25	17.5	4.5	4	2	1			
	4	70	52	9.5	5	2	3			
Mean		48.75	34.62	7.25						

Bark texture

1 - Smooth

2 - Moderately smooth

3 - Moderately rough

4 - R-Rough

A total of 25 plus trees were selected by comparing nearly 102 candidate / comparison trees

GENETIC BASIS AND METHODS OF SELECTION

There are five different basis and methods of selection followed in most tree improvement programme

i. Mass selection
ii. Family selection
iii. Within family selection
iv. Family plus within family selection
v. Selection of parent based on Progeny test

i. Mass selection

It is a method of selection wherein the individuals are selected based on their phenotypic superiority. It is the most commonest methods of individual tree selection used predominantly in improvement of several tree species particularly when the parentage is not known.

ii. Family selection

It is another method of selection wherein the entire family is selected based on their average phenotypic values. Seed sources selection are very good example of family selection.

iii. Within family selection

It is a method of selection wherein the individuals are selected based on their superiority over the family means.

iv. Family plus within family selection

It is a two stage method of selection wherein the families are selected in the first stage followed by selection of individuals within families in the second stage.

v. Selection of parent based on Progeny test

The most precise method of selection is based on progeny testing which involve selection of parents based upon the performance of their progeny.

OTHER METHODS OF SELECTION

i. Tandem selection

In this selection, one desirable trait is selected and breeding proceeded to get a desired improvement. After getting desired improvement in the trait concerned, selection and breeding efforts are concentrated for the other traits.

E.g. Improvement in cellulose content

Reduction in lignin content

Pest and disease resistance

ii. Independent culling

It is a method of multi-trait selection where in the minimum values for each trait of interest is set first. Individuals falls below this standard is removed and individuals falling above this value is retained for breeding program.

iii. Selection index

It is a form of multi-trait selection which combines information on all traits of interest into a single index. Selection index combines family plus individual information for all traits into one index.

iv. Recurrent selection

It is one of the important methods of selection wherein the individuals are selected through many cycles / generations of breeding programme. However it is a long term process wherein the breeder has to wait for several generations to achieve the desired goal.

3. HYBRIDIZATION

Hybridization remains an important method of crop breeding programme which has led to quality and quantity improvement in several agricultural crops. Wide range of hybridization methods have been deployed in agricultural crops. This is due to the existence of homozygosity and self pollination nature of most crops.

However trees species in nature are highly cross-pollinated and the resultant progenies are heterogeneous and heterozygous which are exploited through selection followed by exploitation through progeny and clonal propagation. Hence hybrid breeding programme in tree crop is very minimal and confined only to few species like eucalyptus, casuarina, populus, pines, salix etc., The hybrid breeding methods followed in tree crops are summarized below:

i. Intraspecific hybridization: In this method, the best parents of same species are crossed to obtain productivity improvement or wood quality changes or resistance towards biotic and abiotic stresses.

Ex.

Eucalyptus tereticornis × *Eucalyptus tereticornis*
(ET 3- High Yielder) (ET 103-High cellulose)

↓

F_1 – Hybrid with high yielding and high cellulose characters

ii. Interspecific Hybridization: In this method, better parents of two different species are crossed to obtain yield improvement or wood quality improvement or resistance toward biotic and abiotic stresses etc.,

Ex. i

Casuarina equisetifolia × *Casuarina junghuhniana*
(Clone MTP 12) High Yielder (CJ 27)

↓

F_1 evaluated for yield

↓

High yielding hybrid identified (Casuarina hybrid MTP 2)

Ex. ii

Melia dubia (MTP 2)	*Melia composite* (PAU)
(High Yielder with low wood density)	(Medium wood density with low yield)

↓

F_1 – High yielder with medium density (Under evaluation)

iii. Intergeneric Hybridization: It is a method of hybrid breeding were two different genes are crossed and aimed for achieving specific hybrid with special characters. However intergeneric hybridization programme is very minimally attempted and not much success has been evidenced.

Procedure of hybridization

Once the breeder has decided the objectives of programme, he is ready to begin hybridization. There are seven steps involved in hybridization: (1) choice of parents, (2) evaluation of parents, (3) emasculation (4) bagging (5) tagging (6) pollination, and (7) harvesting and storage of F_1 seed.

i. Choice of parents

The choice of parents mainly depends upon the breeding objective. In addition to other objectives, increased yields are always an objective of the breeder. Therefore, at least one of the parents involved in a cross should be a well adapted variety in the area for which the new variety is being developed.

ii. Emasculation

The removal of stamens or anthers or killing of pollen grains of a flower without affecting in any way the female reproductive organs is known as emasculation. The main objective of emasculation is to avoid self fertilization.

iii. Bagging

After emasculation, the emasculated flower bud is covered suitably to avoid cross pollination from foreign pollens. Generally bagging is done by using butter paper cover.

iv. Pollination

It involves collecting pollen from freshly dehisced anthers of the male parent and dusting this pollen on to the stigmas of the emasculated flowers.

v. Tagging

The emasculated flowers are tagged after bagging indicating the date of emasculation, date of pollination and the names of male and female parents. In general, the name of female parent is indicated first and the male parent is returned later.

For eg. For the cross Eucalyptus MTP 1 × Eucalyptus MTP 7, MTP 1 is the female parent and MTP 7 is the male parent.

vi. Seed harvesting

After seed set, the fruits are collected and seeds are separated out and used for germination.

vii. Raising of F_1 seedlings

The F_1 seeds are sown directly in the polythene bag or sown on the mother bed and the germinated seedlings are pricked out and planted in the polythene containers. Three to six months old seedlings are used for assessment or evaluation at nursery stage. In recent times, the F_1 seeds are germinated directly in the sand bed medium used in mother garden establishment. After germination, the early superior seedling are identified and multiplied through mini clonal technology for further testing as a hybrid clones.

viii. Hybrid evaluation

The F_1 progenies are evaluated in the field conditions along with the check parent for comparisons and to estimate heterosis or hybrid vigour.

4. MUTATION

It is one of the important and special breeding methods used in several crop plants but its application in forestry is very limited. However the breeding method is briefed to benefit the tree breeders of the country.

Mutation is the sudden heritable change in the genetic constitution that occurs naturally at a very low rate or induced artificially by wide range of physical and chemical mutagens.

Mutagens

An agent which is a physical or chemical form involved in changes of genetic constitution is termed as mutagens.

Physical mutagens : Electromagnetic Radiation such as X-rays, gamma rays and UV light. It also includes particle radiation such as fast and thermol neutrons, Beta and Alpha particles.

Chemical mutagens : EMS – Ethyl Methane Sulfanate, Colchicine, Sodium azide, Di-Ethyl Sulphate.

Application of Mutation breeding

- Mutation helps to improve both oilgogenic and polygenic characters
- It helps to alter wood quality.
- In TBOs, this breeding methods will help to change the oil quality.
- Mutation breeding may help to develop thornless varieties particularly in trees like prosopis, ċesalpinia, Inga etc.,
- It is useful to develop traits of specific interest.
- It may also lead to development of novel mutant varieties.

Limitations of Mutation breeding

- One of the major limitations of this method is low frequency of the desired mutations.
- It involves deploying wide range of populations.
- There are chances of chromosomal aberrations which may alter the desired traits.
- Mutant may produce pleiotropic effects in tree crops which may delay flowering.

5. BIOTECHNOLOGY BASED BREEDING

There is a growing interest towards deploying biotechnology and molecular tools in tree breeding and improvement programme. Among several tools and methods, the role of protoplast fusion for production of somatic hybrid and the genetic engineering towards development of Genetically Modified trees are very well witnessed. Successful development of hybrids using protoplast fusion have been reported in species like, pinus, poplus and larix (Puite, 1992), but still the application of somatic hybrids is only of academic interest and very limited attempts were made for commercial exploitation in tree crops due to poor regenetration of plants from protoplast and incompatibility.

Alternatively genetic engineering has been establishing as a very strong tool in tree improvement. Several GMOs were already successfully reported in wide range of tree crops like poplars and eucalyptus. The GMOs tree exhibited increased growth rate, herbicide tolerance, biotic and abiotic resistance coupled with wood quality improvement which have attracted both the researcher and wood based industries towards deploying the technology. However unlike crop plants, the application of genetic engineering technology is confined only to a few species like poplars, pines, eucalyptus, alder, betula, casuarina etc.,

SUMMARY

Compared to agriculture and crop breeding, tree breeding is of recent origin which dates back to 6 to 7 decades old. The traditional tree breeding methods concentrated purely on selection followed by progeny evaluation which has taken long period coupled with huge expenditure. However with the advancement in technologies like Mini Clonal technology and biotechnology tools along with changes in breeding approaches have made tree breeding as a very easy and possible method. The quality characterization of parental trees coupled with exploitation through clonal technology has eliminated the stage of progeny evolution and reduced the breeding period. Similarly the introduction and implementation of controlled breeding have also made tree breeding as an easily approachable process. Considering all these development into account, the practising traditional tree breeding methods along with advances made in breeding procedures have also been incorporated in this chapter with suitable citations and practical demonstration which may find greater utility for the graduating scholars and researchers in tree breeding.

5

HYBRID DEVELOPMENT AND HETEROSIS

INTRODUCTION

Due to increasing wood demand, decreasing supply from natural forest and the restriction on felling of trees in the natural forest have necessitated intensive tree breeding programme both through selection and also by hybrid development programme. The forestry species are highly heterozygous and exhibit wide range of variability in the existing population and hence the need for hybridization and the associated hybrid development program is dismally modest. This hybridization has originally been conducted in temperate species which has now been expanded to almost several tropical species. Successful hybrids in eucalyptus have witnessed yield and wood quality improvements. The increased in yield particularly in the hybrid of *Eucalyptus urograndis* which was bred between *Eucalyptus grandis* and *Eucalyptus urophylla*. Similarly successful hybrid development programme is witnessed in Populus, Pines, Pseudosuga, Picea, Casuarina, Jatropha and other commercially important tree species. As the demand for wood is increasing along with increasing interest on plantation development both within and outside the country, the demand for superior genetic resources development is also as the risk. Under such circumstances hybrid breeding and the associated heterosis development in forestry sector is increasing for yield improvement, wood quality modification and resistance towards biotic and abiotic stresses. This chapter elaborates the hybrid development programme along with heterosis to give basic understanding on the tree breeding process and the associated hybrid vigour development.

DEFINITION OF TERMINOLOGIES

Hybrid	:	The progeny of a cross between genetically dissimilar parents.
Heterosis	:	The superiority of F_1 in one or more characters over its parent.
Intraspecific Hybridization	:	A cross between two varieties or genotypes of the same species.
Interspecific Hybridization	:	Crossing or mating between two different species of the same genus is referred as Interspecific Hybridization.
Intergeneric Hybridization	:	Intergeneric Hybridization refers to crossing between two different genera of the same family.
Hybrid vigor	:	Hybrid vigor is the improved activity and survival of hybrid offspring.
Inbreeding	:	Mating between closely related individuals such as selfing and sib mating.
Inbreeding Depression	:	Loss or decrease in vigour and fitness as a result of inbreeding.

SIGNIFICANCE OF HYBRIDS

Hybrid is the resultant product of crosses between two dissimilar parents. The hybrid development programme is normally done through inherent characters of a parent deployed for a crossing programme in order to increase the yield, form of trees, wood quality, pest and disease resistance, adaptability etc., Hence the tree breeder carefully choose the parents with a desired characteristics. Before initiating any crossing programme, the breeder should have thorough understanding on the idea about the purpose of breeding based on the objectives envisaged. Accordingly the parents can be selected and deployed in hybrid development programme.

TYPES OF HYBRIDS

In trees, generally two types of hybrid development programme are followed viz.,

i. Intraspecific hybrids

ii. Interspecific hybrids

i. Intraspecific hybrids

It is a type of hybrid carried out between genetically different genotypes of the same species. Eg.

Casuarina equisetifolia × *Casuarina equisetifolia*
(Clone MTP 12) (Clone MTP 7)

↓

F_1
(Intraspecific hybrid)

ii. Interspecific hybrids

The hybrid obtained from crosses between two different species of the same genus is termed as inter specific hybrids. Eg.

Casuarina equisetifolia × *Casuarina junghuhniana*
(Clone MTP 12) *(CJ 27)*

↓

F_1
(Interspecific hybrid)

HYBRID DEVELOPMENT PROCESS IN TREE CROPS

Hybrid development programme in tree species is confined only to a very few species of commercial importance; otherwise in most tree species, the tree breeding programme is concentrated primarily through selection followed by evaluation and identification of superior genetic stocks. Only few species are attempted through systematic hybrid development programme and the major hybrid development process approached in various tree species are furnished below. Though hybrid development varies between species, the common procedures followed for most tree species are furnished below.

Identification of potential and superior genetic stock

↓

Raising of hybrid population through clonal technology (Rooting of cuttings or grafting)

↓

Induction of flowers through application of growth regulators and or paclobutrazol

↓

Carried out controlled crosses after flower induction

↓

Production of F_1 seeds (Intraspecific or interspecific)
Raising of F_1 progenies

↓

Identification of hybrid vigor through yield and good quality evaluation

↓

Fixing the hybrid vigor through clonal technology

↓

Mass multiplication of the identified heterotic progeny through mini clonal technology

↓

Evaluation through Multi Location Trials (MLT)

↓

Promotion for adoption

The following are the major hybrid development programme carried out at Forest College and Research Institute for varied purposes and objectives.

i. Development of High yielder in eucalyptus

Eucalyptus tereticornis × *Eucalyptus camaldulensis*

↓

F_1

↓

F_1 Population evaluated

↓

Identification of superior Hybrid progenis

↓

Exploitation through clonal technology

↓

Mass multiplication and deployed in operational forestry programme

ii. Development of high cellulose eucalyptus

Eucalyptus tereticornis × *Eucalyptus camaldulensis* (EC 48)

↓

F_1

↓

F_1 Population evaluated for wood quality

↓

Elite Hybrid clone with high cellulose content

iii. High yielding varieties in casuarina

Casuarina equisetifolia (Clone MTP 12) × *Casuarina junghuhniana* (CJ 27)

↓

F_1 evaluated for yield

↓

High yielding hybrid identified (Casuarina hybrid MTP 2)

iv. Development of semi hardwood stem based hybrid in Jatropha

Jatropha curcas (TNMC 7) × *Jatropha integerrima* (Wild with semi hard stem)

↓

F_1

↓

Hybrid with semi hardwood identified but with poor seed yield

CONTROLLED BREEDING IN TREE CROPS

In the recent past, the traditional hybridization programme of directly crossing on the mature tree is replaced with an advanced control breeding methods. This method has been followed intensively in Brazil pulpwood plantation development program which led to development of superior hybrids with higher productivity, improved wood quality and reduction in rotation age. This control breeding has been followed for eucalyptus hybrid development programme which has great potential for replication in other species. Such control breeding has also been applied in casuarina at FC&RI campus, Mettupalayam, India which results in successful development of inter-specific hybrids between *Casuarina equisetifolia* and *Casuarina junghuhniana*. Currently this hybrid technology is being applied for *Melia dubia* and *Melia composita*. The control breeding method involved the following strategies

i. **Development of grafted plants**

The control breeding method involve development and establishment of breeding orchard deploying clones propagated through cleft grafting technology which involve stocks raised from seedling and scions obtained from identified plus trees. The scions are inserted on the stocks and grafted for successful union. The grafted plants are then used for early flower induction both through physical and chemical methods. The methods followed for eucalyptus and casuarina breeding programme are furnished below

ii. ***In-situ* control breeding**

- In this method, the grafted plants are planted in the breeding field orchard at an espacement ranging from 2 to 4m. Then the plants are treated both physically and chemically in order to induce early flowering.
- The physical treatment involves twisting the partially mature branches which creates stress to induce flowering. This can be done manually by hands or by using flexible wires.
- Secondly the plants are treated with paclobutrazol as soil application around the root zone of the plant.
- These two treatments facilitate early flower induction after six months to one year of treatment.

- The leading apical shoots are cut at 1.5 m to 2 m height to facilitate induction of more branches and to induce more flowers at reachable heights.
- In case of eucalyptus after the occurrence of flower, the mature flower buds (Prior to anthesis) are selected and the apical dome of the flower bud is cut using scissors.
- Once the stigma is visualized, the pollen collected from identified superior genetic stocks is dusted directly on the stigma.
- This facilitates successful fertility and development of hybrids
- The crossed F_1 seeds are collected after maturity preferably during the period between 4 and 6 months.
- Then the seeds are sown in the mother bed and seedlings are raised.
- In the hybrid seedling population, elite seedling are selected and mass multiplied through clonal multiplication technology and deployed as hybrid clones for direct evaluation
- Alternatively the F_1 progenies are evaluated and among them superior clones are identified based on yield and wood quality which are further mass multiplied and deployed in operational clonal plantation establishment.

iii. *Ex-situ* control breeding

- In this method also the same methodology is followed. However the grafted plants instead of planting in the field are grown in big plastic container or cement pot containers.
- The grafted plants allow to grow for 6 months to 1 year till it reaches an acceptable height
- After that cutting of leading shoot, application of physical and chemical methods to induce flowering, crossing programme, production of F_1 hybrids and deployment of clonal technology followed by clonal evaluation and utilization through operational plantation development are followed as per the method described earlier.

Controlled breeding Chamber

NATURAL HYBRIDS

Tree species by virtue of their stature, distribution and interaction within the natural forest ecosystem and in certain cases of established plantation, natural hybridization occurs spontaneously which lead to development of inter and intra specific hybrids. Very classical example is natural inter-specific hybrids between *Casuarina equisetifolia* and *Casuarina junghuhniana* leads to development of a sterile hybrid. This sterile hybrid has been identified as most productive both in natural ranges and also in the introduced locations.

Similarly, identification of spontaneous hybrids in eucalyptus plantation of Brazil is yet another example which led to productivity improvement in Brazil. The natural hybrids and associated hybrid vigour has been exploited through advanced clonal technology in Brazil.

In maple also, a natural inter-specific hybrids has been reported and it's deployed for commercial cultivation. Sometimes the population which grows together freely hybridize and further involved in backcross breeding which occurs naturally. The population obtained from a natural hybrid followed by natural backcross is pronounced as hybrid swamps. The natural hybrids and hybrid swamps have been reported in several species of pines, junipers, oak and also in eucalyptus (Mandal and Gibson, 1998; Zobel and Talbert. 1984).

Successful Commercial Hybrids in trees

S.No.	Species	Reference
1.	*Eucalyptus grandis* × *E. urophylla* *Eucalyptus urograndis*	Parthiban, 2017
2.	*Salix alba* × *S. matsudana*	Bhojwani, 1980
3.	*Casuarina equisetifolia* × *C. junghuhniana*	Parthiban *et al.*, 2018
4.	*Jatropha curcas* × *J. integerrima*	Parthiban *et al.*, 2009
5.	*Picea sitchensis* × *P. glauca*	Stockes *et al.*, 2018
6.	*Populus deltoides* × *P. euphratica*	Singh *et al.*, 2003
7.	*Pinus elliottii* × *P. caribaea*	Dieters and Brawner, 2007
8.	*Populus nigra* × *P. x canadensis*	Smulders *et al.* 2008
9.	*Acacia mangium* × *A. auriculiformis*	Kha, 1996
10.	*Quercus ilex* × *Q. suber*	Soto *et al.*, 2003

COMBINING ABILITY

- It is the ability of parents or crosses to combine among each other during hybridization process which results in transmission of desirable genes or characters to their progenies.
- It is also an estimation of the value of genotype on the basis of their offspring performance in a definite mating design.
- The crosses between parents result in production of a potential offspring is also said to be combining ability.

Types of combining ability

In general, two types of combining ability are classified

i. General Combining Ability (GCA)

It is the average performance of a genotype in a series of hybrid combinations.

ii. Specific Combining Ability (SCA)

It is a combining ability in which certain hybrid combination performs better or poorer than the expected levels on the basis of average performance of the parental inbred lines.

Comparison between GCA and SCA

S.No	GCA	SCA
1.	GCA is an average performance of a genotype in a series of crosses	It is the performance of specific cross in relation to the average performance to the parental inbred lines.
2.	It is attributable through half sib families where in one parent is known	It is attributable due to full sib families
3.	Selection of suitable parents for hybridization is amenable through GCA	SCA helps in identification of superior cross combination
4.	GCA has relationship with narrow sense heritability	SCA has relationship with heterosis
5.	The GCA is attributable due to additive genetic variance and additive × additive epitasis	The SCA is attributable due to dominance genetic variance and all three types of epitasis

Application of Combining Ability in trees

- It helps to increase the biomass yield as most yield components are of polygenic control.
- Good Combining Ability will help to improve the wood quality in terms of density, cellulose, hemicelluloses and lignin composition.
- In case of tree borne oil seeds the combining ability will help to improve the oil quality
- Several tree species are known to contain antioxidant properties and the good combining ability may help to improve the antioxidant properties
- The combining ability is also helpful in tree breeding to develop varieties for pest and disease resistance.

HETEROSIS

The hybrid development and associated hybrid vigour is inseparable because the ultimate aim of hybrid development is to ensure hybrid vigour which is often pronounced as heterosis. Heterosis is the process in which, the F_1 population obtained through the crosses between two genetically dissimilar parents lead to increased or decreased vigour over the better parent or over the mid parent value. The scientist Shull has first coined the word "Heterosis" and the interpreted the word heterosis in terms of increase in vigour, size, fruit fullness, rapid development of resistance to pest and disease compared to the parents or the corresponding inbreeds. Today the hybrid vigour and heterosis and its application has widely been exploited in agricultural sector and is receiving significant attraction from the tree breeders. Unlike agriculture crops, the heterosis in tree crops are towards yield and wood density, low bark, high cellulose, low lignin, pest and disease tolerance, herbicide tolerance, Saline and alkaline tolerance etc., Today wide ranges of tree species have been attended for hybrid development programme and the associated heterosis. This heterosis has been classified into various types as based on

a. Function

b. Transmissibility

c. Commercial utility

A. Types of heterosis based on function

Based on the function, the heterosis is classified into the following types

i. Luxuriant Heterosis

It is a type of heterosis, wherein the evaluation is restricted to one or few characteristics which mean that one trait may have positive luxuriant and another trait may have negative luxuriant.

ii. Adoptive Heterosis

It is a type of heterosis, where the evaluation is based on performance or fitness of hybrids to the environment.

iii. Selective Heterosis

In this case, the evaluation based on advantages in vegetative phase and not in the generative phase.

iv. Reproductive Heterosis

It is the type of heterosis wherein the hybrids are evaluated based on productivity.

B. Types of heterosis based on transmissibility

There are 2 types of heterosis based on Transmissibility viz., liable heterosis and fixed heterosis

i. Liable heterosis

It is the type of heterosis which desiccated due to frequent segregation heterosis in the next sexual generations. It includes heterozygous heterosis in higher plants and heterokaryotic heterosis in lower plants.

ii. Fixed heterosis

It is the type of heterosis wherein the heterosis is fixable in balanced heterozygous or homologous and homozygous conditions.

C. HETEROSIS BASED ON COMMERCIAL IMPORTANCE

Among all the types of heterosis, the heterosis classification based on commercial utility is of great significance. Based on commercial utility, the heterosis is classified into three types viz., relative heterosis, heterobeltiosis and standard heterosis which are discussed below.

i. Relative Heterosis

It is the type of heterosis which is defined as the superiority of hybrid over mean parent value

$$Relative\ heterosis = \frac{F_1 - Mean\ parent\ value}{Mean\ parent\ value} \times 100$$

ii. Heterobeltiosis

It is the type of heterosis which is defined as the superiority of hybrid over the better parent

$$Heterobeltiosis = \frac{F_1 - Better\ parent}{Better\ parent} \times 100$$

iii. Standard Heterosis

It is the type of heterosis wherein the superiority of hybrid over the commercial check variety is ensured. This heterosis has practical significance in tree breeding and is otherwise referred as useful heterosis or economic heterosis.

$$\textit{Standard heterosis} = \frac{F_1 - \text{Check variety}}{\text{Check variety}} \times 100$$

Based on increasing or decreasing value with respect to the trait of interest, the heterosis is again classified into two types viz., positive and negative heterosis

i. **Positive Heterosis**

If the F_1 hybrid expresses increased vigour over its parent it is called positive heterosis

ii. **Negative Heterosis**

In this case, the F_1 hybrid expresses decreased vigour over its parent.

PRACTICAL DEMONSTRATION OF HETEROSIS ESTIMATION

The productivity of *Casuarina equisetifolia* and *Casuarina junghuhniana* are 80 tonnes / ha and 100 tonnes / ha respectively. The productivity of F_1 hybrid between these two species is 150 tonnes compared to the commercial check variety yields of 100 tonnes / ha. Calculate the heterosis based on commercial utility.

Solution

- The productivity of *Casuarina equisetifolia* = 80 tonnes / ha
- The productivity of *Casuarina junghuhniana* = 100 tonnes / ha
- The productivity of F_1 hybrid = 150 tonnes / ha
- The productivity of check variety = 100 tonnes / ha

i. **Relative Heterosis**

$$\textit{Relative heterosis} = \frac{F_1 - \textit{Mean parent value}}{\textit{Mean parent value}} \times \textit{100}$$

$$= \frac{150 - 90}{90} \times 100 = 0.6667 \times 100$$

$$= 66.67$$

ii. **Heterobeltiosis**

$$\textit{Heterobeltiosis} = \frac{F_1 - \text{Better parent}}{\text{Better parent}} \times 100$$

$$= \frac{150-100}{90} \times 100 = 0.50 \times 100$$

$$= 50.00$$

iii. Standard Heterosis

$$\textit{Standard heterosis} = \frac{F_1 - \text{Check variety}}{\text{Check variety}} \times 100$$

$$= \frac{150-100}{100} \times 100 = 0.50 \times 100$$

$$= 50.00$$

Conclusion

The heterosis estimation indicates that the developed hybrids are highly superior compared to the mid parent, better parent and also the standard check variety. Considering all the values and giving priority to standard heterosis, the inter-specific hybrids established between *Casuarina equisetifolia* and *Casuarina junghuhniana* have got excellent hybrid vigour and hence these hybrids could be deployed in operational plantation establishment for commercial and industrial utility.

GENETIC BASIS OF HETEROSIS

The genetic basis of heterosis is discussed elaborately by wide range of researchers but the major effect of heterosis has been explained through two major theories

i. Dominance hypothesis

ii. Over dominance hypothesis

Before discussing the hypothesis, it is also very significant to discuss the theory of inbreeding depression which means that the lowered fitness or vigour of inbred individuals compared to their non inbred counterparts.

Generally the inbreeding depression and heterosis are opposite to each other wherein heterosis relates to increased growth rate and vigour and inbreeding depression relates to decreased vigour. However these two principles are primarily due to non additive gene action and or considered to be of two different aspects of the same phenomena.

i. **Dominance hypothesis**

The theory of dominance hypothesis was conceived by Devanport (1908) and followed by several researchers like Bruce, Keeble and Pellew. This theory of dominant hypothesis indicates that at each locus, the dominant allele has a favourable effect and the deleterious effects of recessive alleles are masked by their dominant alleles. It reveals that the heterosis effect is due to masking effect of dominant alleles. In case of inbreeding depression, the expressions are due to harmful effects of recessive alleles which become homozygous due to inbreeding. Hence it is very clear that as per the theory of dominant hypothesis, the expression of heterozygosity effect of heterosis is not due to heterozygosity but due to prevention of expression of harmful recessives by their corresponding dominant alleles. Similarly expression of inbreeding depression is not due to homozygosity but due to the homozygosity of recessive allele's accredited with harmful effects.

ii. **Over dominance hypothesis**

The principle behind the over dominance hypothesis was proposed independently by East and Shull (1908) and the terminology of over dominance was coined by Shull in the year 1945. This theory explains that heterosis is due to heterozygosity which is essential and acts as a major cause for hybrid vigour.

FIXATION OF HETEROSIS IN TREES

Trees unlike annual crops demand a specific approach to fix and exploit the heterosis approach to fix and exploit the heterosis or hybrid vigour. However in practise, there are four different ways of fixation of heterosis.

1. Vegetative propagation
2. Apomixis
3. Balanced lethal system
4. Polyploidy

Among all these process, the heterosis in tree crops are exploited predominantly by clonal technology through vegetative propagation. In this case, the F_1 hybrids based on superiority of yield and wood quality, harvested, coppice shoots are induced and these coppice shoots are multiplied through miniclonal technology. This technology results in mass multiplication of hybrid clones and deployed in evaluation and other operational plantation development program.

APPLICATION OF HETEROSIS

Heterosis may lead to increased growth and development and the advantages of heterosis in the crop are furnished below

1. **Productivity Improvement**

   One of the major advantages of heterosis is yield or productivity improvement in tree crops. This is very clearly evidenced in *Eucalyptus urograndis* in Brazil which exhibited increased yield of over 40m^3/ ha/ annum due to the exploitation of heterosis and hybrid vigour.

2. **Improvement in general size and vigor**

   Heterosis and hybrid vigour has created healthier and faster growth in several tree species like Casuarina, Subabul, Eucalyptus and Jatropha whish lead to increased vigour and luxuriant growth.

3. **Improved wood quality**

   There are evidences in Eucalyptus and Pines wherein the F_1 progenies exhibited improved wood quality in terms of higher cellulose and lower lignin due to heterosis effect.

4. **Wider adaptability**

   The hybrids developed in trees have exhibited wider adaptability compared to their respective parents.

5. **Increased biotic and abiotic resistance**

   Heterosis may also lead to increased resistance against both biotic and abiotic components due to the development of pest and disease variety as witnessed in Poplar and saline and alkaline tolerance as witnessed in Casuarina.

6. **Higher reproductive ability**

   The heterotic effect of hybrids has expressed increased fertility rate and contribute towards enhanced reproduction.

7. **Early maturity**

   Some of the hybrids in tree crops particularly Eucalyptus and Casuarina have been exploited due to their rapid growth and short duration harvest which benefited the growers towards early returns and wood based industries through the sustained development of raw material.

8. Above all, the heterosis and the associated hybrid vigour have manifested and played a significant role at molecular, functional, cellular and also at organism levels.

SUMMARY

Hybrid development is a fundamental process for productivity and wood quality improvement. Traditional tree breeding approaches are contributing primarily in the selection from the existing base population which is highly heterogenous and heterozygous. However in certain species, the existing variability may be narrow which demand induced variation through hybridization or mutation based breeding approaches. Among these two approaches, the potential of hybridization and the associated hybrid vigour is very well witnessed. Successful hybrids are now available in wide range of species like Eucalyptus, Casuarina, Poplars, Pines etc., with an increased yield potential and wood quality improvement. The superiority of hybrid over parent is very well witnessed in *Eucalyptus urograndis,* which exhibited productive potential of over 40m^3/ ha/ annum. Similarly existence of hybrid vigour in Casuarina, Jatropha and Pines are almost established for increased yield. The genetic basis of hybrid development and behavior due to dominance and over dominance hypothesis is very well discussed for easy understanding. This besides, the estimation of heterosis and various types of heterosis are demonstrated with suitable examples and model calculations. The hybrid development process, successful hybrids in various species, hybrid vigour and its advantages are also provided in this chapter to benefit graduating professionals of forestry and other tree breeders for practical application in tree breeding program. Above all, the chapter will find greater utility to meet the needs of competitive examinations.

6

GENETIC TESTS AND TOOLS

INTRODUCTION

Tree breeding and improvement has been recognized as a long term approach traditionally and take several years to obtain the genetic gain. This was considered as major obstacles in tree breeding programme. However with the advances in technology and the changes in breeding approach particularly in tropical tree species have made tree breeding as a short duration project and help to achieve the desired results in a shorter period. This tree breeding has gained significant importance due to growing demand for wood and wood products and the restriction on felling of wood from the natural forest. The changes in policy and legal issues have necessitated the promotion of exclusive industrial plantations and agroforestry systems across the country to create self reliance in raw material security. These plantation programme requires improved genetic resources with higher productive potential and improved wood quality to cater to the needs of varied industrial utility besides acting as a resources for domestic wood needs. The improvement of tree genetic resources is carried out through various approaches like provenances test, seed sources evaluation, progeny and clonal tests. The superior genetic stocks identified and screened from these standard breeding tests are deployed in operational forest plantation development programme. This chapter deals with various genetic test viz., provenance, seed source, progeny and clonal test deployed in tree breeding and improvement programme with suitable field level applications to benefit graduates, professional foresters and general readers of the book.

DEFINITION OF TERMINOLOGIES

Provenances	:	The original geographic area from which seed or other propagules were obtained. It is also known as Geographical source or geographical race
Seed sources	:	Location of the parent trees from which seed was used for establishing a plantation.
Progeny testing	:	The best way to evaluate the genetic worth of selected parents is to grow their progeny in a way that allows estimation of the parental breeding value.
Clone	:	The aggregate of individual organisms, originating from one sexually produced individual and maintained exclusively by asexual propagation. A group of vegetatively-propagated organisms consisting of an ortet and its ramets. A cell line of single-cell origin.
Ecotype	:	An ecotype is a group of plants of similar genotype that occupy a specific ecological niche. Ecotype is a genotypical response of a species to a particular habitat.
Cline	:	Cline is a gradient in a measurable characteristic. Clines refer to a given characteristic and not to the entire genetic constitution of a population.
Candidate Plus Tree	:	A tree that has been selected for grading because of its desirable phenotypic qualities but that has not yet been tested or graded
Plus Tree	:	A tree that has been recommended for production or breeding orchard use following grading. It has a superior phenotype for growth, form, wood quality or other desired characteristics and appears to be adaptable. It has not yet been tested for its genetic worth, although the chances of having a good genotype are high for characteristics with a reasonable heritability.

Elite Tree : Trees that have proven to be genetically superior by means of progeny testing.

Winner from a selection program & is the kind of tree that is most desired for mass production of seeds or vegetative propagules

Comparison Tree : Trees that are located in the same stand are of nearly the same age, are growing on the same sites as the select tree & against which select tree is graded.

GENETIC TESTS

The following are the major genetic test practically adopted in all tree breeding programme towards identifying superior genetic resources for deploying them in operational forestry programme besides acting as a tool for conservation of Forest Genetic Resources.

I. Provenance Test

II. Seed Sources Evaluation Test

III. Progeny Test (PT)

IV. Clonal Test (CT)

I. PROVENANCE TESTS

Provenance means original geographical area from where the seeds or other propagules are collected. The provenance test means evaluation of different provenances of a species in order to estimate the probability that any better genes or populations can be found atleast for one trait that can be used to improve the productivity of a plantation.

Seed collection for provenance trial

- First phase of provenance test is to identify the range of provenances based on the existing knowledge and the variability that exists in the species. It is always better to deploy a minimum of 15 provenances in a species and a maximum of 30 provenances depending on the existing variability.
- Once the provenance are identified based on the natural variability, then collection of seeds from the identified provenances are very important.

- Several authors suggested collection of seeds from 20-30 trees per provenance and mix them for use as a single provenances.
- Many international provenance tests indicated that bulk seed lot composed of seeds collected from many trees has also been recommended
- However the ideal provenance seed collection is within the provenance. Select 10-30 trees based on morphometric attributes like height, GBH, Tree form, free from pest and diseases etc., Seeds can be collected from these identified trees and deployed as a provenance.
- While selecting provenances care should be taken that edapho-climatic conditions are taken into account for evaluation. Introduction of provenance in contrasting sites and climate need not be carefully examine and accordingly the provenances are to be introduced and evaluated.
- Higher altitude provenances should not be selected for planting in the lower altitude and vice versa
- Similarly coastal provenances need not be selected for planting in Inland conditions.

Plot size

- The provenance test plots are variable due to the number of provenances, objective and number of individuals amenable for evaluation within the provenances. Hence the plots may be of row column or rectangular or square block based experiment plot. For most practical application, the square block of test plot is very effective which will eliminate the competitive effects between the provenances.

Design

- Several statistical designs are useful for provenance testing but the most common and simplest form of design for the provenance test is the Randomized Complete Block Design.
- RBD is very simplest design and the use of replications or blocks will help to separate out undesirable environmental variations and also helps to make difference among provenances.
- It is very easy to layout in the field and increases the statistical accuracy.

Types of Provenance Tests

The provenance tests include range wide provenance phase, restricted provenance phase and the provenance proving phase.

a. **Range wide provenance phase**

Objective: To determine the extent and pattern of variations between provenances of promising species with wide natural variation.

Experimental details

Number of provenances : 10 - 30
Period of evaluation : Quarter to Half rotation (0.25 – 0.5 × rotation age)

b. **Restricted provenance phase**

Objective: To test the suitability of provenances to site under tests

Experimental details

Number of provenances : 3 - 5
Period of evaluation : Half rotation (0.5 × rotation age)

c. **Provenance proving phase**

Objective: To test the genetic worthiness of the identified provenance towards the objectives envisaged (either productivity or resistance or wood quality)

Experimental details

Number of provenances : 1 or 2
Period of evaluation : Half rotation (0.5 × rotation age)
Spacing : 3 m × 3m
Number of trees : 25 (5 × 5 sq. plot)

Simultaneously the general provenance tests can also been carried out through the following standard evaluation tests.

Number of provenances : 15 - 30
Number of replications : 3-10
Spacing : 3m × 3m
Number of trees / plot : 9 or 16 or 25 or 36 or 49 (Depending on the resources available)

Evaluation period : Half to full rotation

PRACTICAL FIELD TEST OF TEAK PROVENENCES TESTS

Name of the Species	–	Teak (***Tectona grandis***)
Name of the Evaluation	–	Provenances test
Location	–	Forest College and Research Institute, Mettupalayam
	–	Latitude: 11°18' N and Longitude: 76°59' E
Breeding Objective	–	To identify the best provenance suitable for semi-arid locality for higher productivity and wood quality
Number of provenances	–	30 (28 Indian provenances and 1 each from Syyaburry and Chittangoong)
Plot size	–	81 m^2
Number of trees / plot	–	9 trees / plot
Design	–	RBD
Number of replication	–	Ten
Method	–	Block plot
Spacing	–	3 m × 3m

Growth attributes evaluated

- Height (m)
- Girth at Breast Height (cm)
- Number of branches
- Volume (m^3)

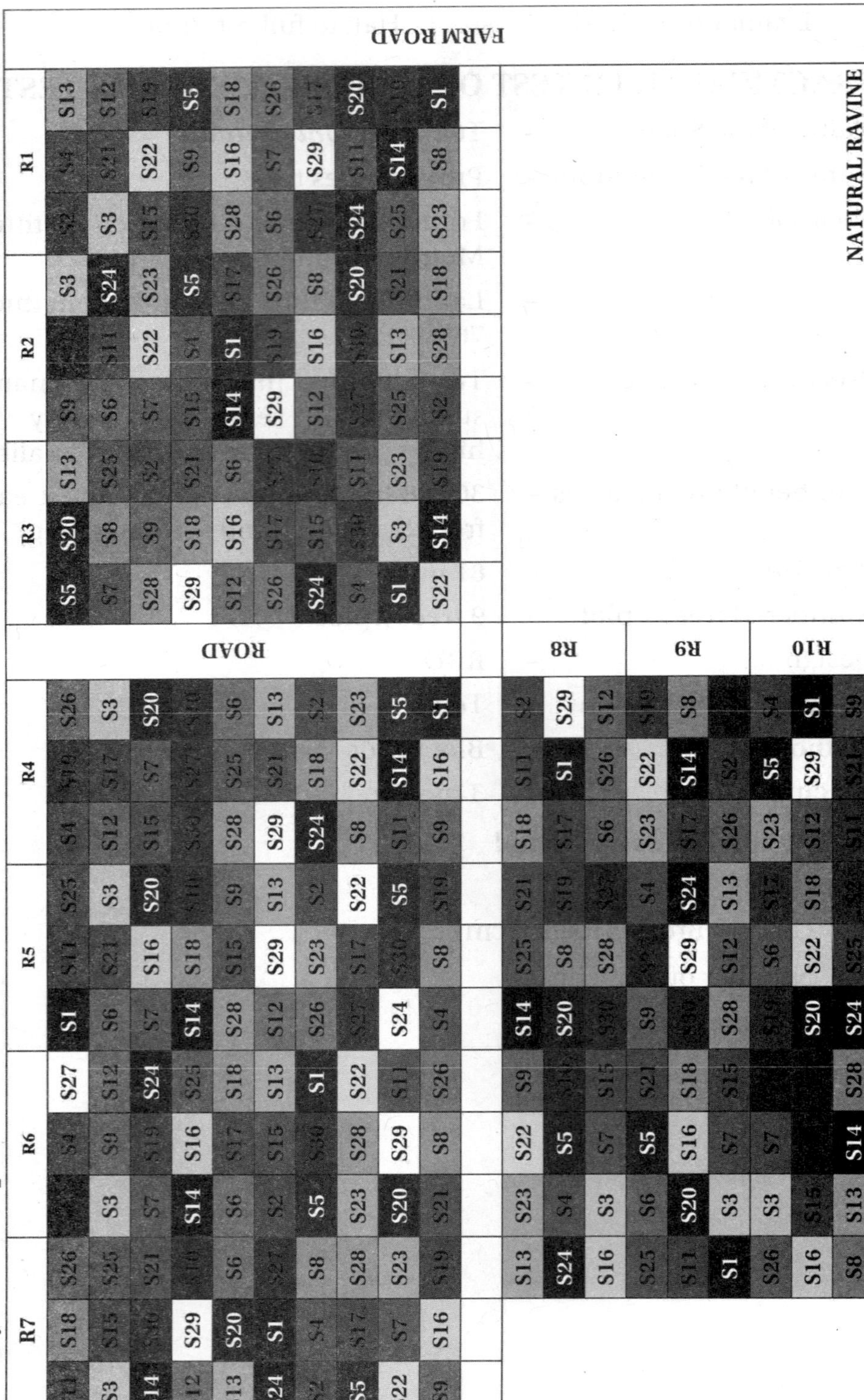

a. Layout of the provenance test

b. Pedigree of the provenance test

S.No.	Provenance	Location	State	Country	Latitude	Longitude
1.	Lao-PDR	Syyabury	-	Lao P.D.R.	23° 31'N	100° 05'E
2.	Bangladesh	Chittagong	-	Bangladesh	22° 19'N	91° 41'E
3.	Uttar Pradesh	Tarai	Uttar Pradesh	India	24° 39'N	78° 12'E
4.	Uttranchal	Dehra Dun	Uttranchal	India	30° 20'N	78° 02'E
5.	Madhya Pradesh	Kottiwara	Madhya Pradesh	India	20° 29'N	74° 08'E
6.	Assam	Lanka	Assam	India	27° 58'N	89° 45'E
7.	Meghalaya	Tura	Meghalaya	India	25° 31'N	90° 13'E
8.	Tripura	Bagafa	Tripura	India	23° 25'N	91° 35'E
9.	Orissa	Kariarroad, Nuapada	Orissa	India	20° 48'N	82° 32'E
10.	Maharashtra-I	Jimalgatt	Maharashtra	India	19° 05'N	80° 10'E
11.	Maharashtra-II	Umari	Maharashtra	India	20° 06'N	78° 58'E
12.	Maharashtra-III	Darekasa	Maharashtra	India	21° 15'N	80° 35'E
13.	Maharashtra-IV	Dongargaon	Maharashtra	India	20° 04'N	80° 17'E
14.	Maharashtra-V	Hiwara	Maharashtra	India	21° 31'N	79° 28'E
15.	Maharashtra-VI	Paoni	Maharashtra	India	20° 30'N	79° 22'E
16.	Andaman	Wimberligunj	Andaman and Nicobar Island	India	11° 45'N	92° 43'E
17.	Karnataka	Sampangi	Karnataka	India	12° 29'N	75° 33'E

S.No.	Provenance	Location	State	Country	Latitude	Longitude
18.	Tamil Nadu-I	Mudmalai	Tamil Nadu	India	11° 37'N	76° 34'E
19.	Tamil Nadu- II	Siruvani	Tamil Nadu	India	10° 55'N	76° 41'E
20.	Tamil Nadu-III	Topslip	Tamil Nadu	India	10° 27'N	76° 50'E
21.	Tamil Nadu-IV	Courtrallum	Tamil Nadu	India	8° 55'N	77° 16'E
22.	Tamil Nadu-V	Kalakkad	Tamil Nadu	India	8° 30'N	77° 32'E
23.	Kerala – I	Wadu-Konni	Kerala	India	9° 13'N	76° 51'E
24.	Kerala – II	Padam-Konni	Kerala	India	9° 13'N	76° 51'E
25.	Kerala – III	Kallaley-Konni	Kerala	India	9° 13'N	76° 51'E
26.	Kerala – IV	Karalai-Nilambur	Kerala	India	11° 16'N	76° 13'E
27.	Kerala – V	Nellikuthu-Nilambur	Kerala	India	11° 16'N	76° 13'E
28.	Kerala – VI	Chatanpura-Nilambur	Kerala	India	11° 16'N	76° 13'E
29.	Kerala - VII	Tholpatty-Wynad	Kerala	India	11° 39'N	76° 16'E
30.	Kerala - VIII	Chettivara-Parambikulam	Kerala	India	10° 23'N	76° 48'E

c. Evaluation of growth attributes of teak provenances (6 MAP)

S. No.	Provenances	Plant Height (cm)	Collar Diameter (cm)	Volume index (cm^3)
1.	Lao-PDR	47.52	1.81	189.69
2.	Bangladesh	47.74	1.76	174.53
3.	Uttar Pradesh	41.91	1.56	120.32
4.	Uttranchal	35.60	1.50	95.08
5.	Madhya Pradesh	20.01	0.93	21.21
6.	Assam	49.45	1.81	179.62
7.	Meghalaya	49.87	1.95	175.82
8.	Tripura	63.88*	2.13*	311.36
9.	Orissa	42.51	1.47	123.74
10.	Maharashtra-I	33.06	1.23	67.16
11.	Maharashtra-II	32.41	1.17	52.47
12.	Maharashtra-III	31.59	1.28	66.34
13.	Maharashtra-IV	38.45	1.38	91.41
14.	Maharashtra-V	41.98	1.50	101.03
15.	Maharashtra-VI	40.17	1.43	87.40
16.	Andamans	57.32	1.88	335.22*
17.	Karnataka	66.29*	2.17*	366.56*
18.	Tamil Nadu -I	56.54	1.83	197.69
19.	Tamil Nadu -II	53.69	1.65	167.28
20.	Tamil Nadu-III	67.21*	2.16*	416.93*
21.	Tamil Nadu-IV	64.07*	1.83	234.35
22.	Tamil Nadu-V	31.73	1.35	61.59
23.	Kerala - I	56.43	1.84	221.09
24.	Kerala - II	52.59	1.75	197.41
25.	Kerala - III	55.57	1.73	187.23
26.	Kerala - IV	52.53	1.78	181.02
27.	Kerala - V	62.36	2.04	312.97
28.	Kerala - VI	68.31*	2.17*	364.58*
29.	Kerala - VII	73.21*	2.27*	408.57*
30.	Kerala - VIII	66.07*	1.90	272.12
	Mean	**49.99**	**1.70**	**192.73**
	SEd	**6.89**	**0.17**	**71.69**
	CD	**13.58**	**0.34**	**141.17**

Inference / Findings

- The provenances differed significantly for various growth attributes like Plant Height, Collar Diameter, No of leaves and Volume index. Since there is no standard variety in teak the mean value has been taken for comparison.
- Compared to the average plant height (49.99 cm), seven provenances viz., Kerala VII (73.21 cm), Kerala VI (68.31 cm) , Tamil Nadu III (67.21 cm), Karnataka (66.29 cm), Kerala VIII (66.07 cm), Tamil Nadu IV (64.07 cm) and Tripura (63.88 cm) exhibited significantly higher plant height.
- The collar diameter of the teak provenances exhibited significant differences. Compared to the average collar diameter (1/70 cm), 5 provenances viz., Kerala VII (2.27 cm), Kerala VI (2.17 cm), Karnataka (2.17 cm), Tamil Nadu III (2.16 cm) and Tripura (2.13 cm) exhibited significantly increased collar diameter.
- Volume index is estimated using height and collar diameter and it differed significantly due to provenances. Compared to the average volume index recorded by all provenance (192.73 cm^3), five provenances viz., Tamil Nadu III (416.93 cm^3), Kerala VII (408.57 cm^3), Karnataka (366.56 cm^3), Kerala VI (364.58 cm^3) and Andamans (335.22 cm^3) registered significantly higher volume index.
- Considering all parameter into account four provenances viz., Kerala VII, Kerala VI, Tamil Nadu III and Karnataka registered significantly increased biometric growth and these provenances are identified with sharp focus in teak improvement programme.

II. SEED SOURCES EVALUATION TEST

Seed sources are the places from where the seeds are collected which may or may not include provenance tests. Very classical example is eucalyptus. If the Lara provenance is introduced to India and if the seeds are collected from plantations grown in India then it is called as Indian seed source and Lara provenances. Unless otherwise the exact provenances are known seed sources from where seeds are collected is pronounced predominantly as seed sources. The evaluation of these seed sources through systematic collection and evaluation is termed as seed sources evaluation tests.

Seed collection for Seed sources evaluation

- The first and foremost step is identification of potential seed sources for the species. The seed sources are identified based on the existing variability coupled with objectives envisaged in a breeding programme of the species.
- Once the seed sources are identified, then collection of seeds from the seed sources is very important.
- For better and efficient seed sources evaluation, it is ideal to identify a minimum of 100 superior trees interms of growth attributes, tree form, pest and disease free and other desirable characters depending on the objectives envisaged.
- From the identified 100 plus trees of a single seed sources, seeds are collected, mixed and the bulked seed lot is termed as a seed source and deployed in evaluation program.

Plot size

- The plot size of seed sources is highly variable like row column or rectangular or square plot based. Among the various plot size, the square plot size is highly efficient which will eliminate the competitive variable between seed sources.

Design

- The most common and simple design deployed for seed source evaluation is randomized complete block design. The design is very easy to establish besides increasing the statistical accuracy if the number of replication is increased.

Practical Field test of Kadam seed sources evaluation test

Name of the Species	–	Kadam (***Neolamarckia cadamba***)
Breeding Objective	–	To screen HYSR seed source
Name of the Evaluation	–	Seed Source Evaluation Test
Location	–	Forest College and Research Institute, Mettupalayam
	–	Latitude: 11°18' N and Longitude: 76°59' E

Number of Seed sources	–	20
Design	–	RBD
Number of replication	–	Three
Method	–	Block plot (9 tree plot)
Spacing	–	3 m × 3m
Period	–	5 years

Growth attributes evaluated

- Height (m)
- Clean Bole Height (m)
- Basal Girth (cm)
- Girth at Breast Height (cm)
- Number of branches
- Volume (m^3)
- Clean Bole Volume (m^3)
- Biomass (kg/tree)

a. Layout of the Seed Sources Evaluation Test (SSET)

R1	AC1	AC1	AC1	AC2	AC2	AC2	AC3	AC3	AC3	AC4	AC4	AC4	AC5	AC5	AC5
	AC1	AC1	AC1	AC2	AC2	AC2	AC3	AC3	AC3	AC4	AC4	AC4	AC5	AC5	AC5
	AC1	AC1	AC1	AC2	AC2	AC2	AC3	AC3	AC3	AC4	AC4	AC4	AC5	AC5	AC5
	AC6	AC6	AC6	AC7	AC7	AC7	AC8	AC8	AC8	AC9	AC9	AC9	AC10	AC10	AC10
	AC6	AC6	AC6	AC7	AC7	AC7	AC8	AC8	AC8	AC9	AC9	AC9	AC10	AC10	AC10
	AC6	AC6	AC6	AC7	AC7	AC7	AC8	AC8	AC8	AC9	AC9	AC9	AC10	AC10	AC10
	AC11	AC11	AC11	AC12	AC12	AC12	AC13	AC13	AC13	AC14	AC14	AC14	AC15	AC15	AC15
	AC11	AC11	AC11	AC12	AC12	AC12	AC13	AC13	AC13	AC14	AC14	AC14	AC15	AC15	AC15
	AC11	AC11	AC11	AC12	AC12	AC12	AC13	AC13	AC13	AC14	AC14	AC14	AC15	AC15	AC15
	AC16	AC16	AC16	AC17	AC17	AC17	AC18	AC18	AC18	AC19	AC19	AC19	AC20	AC20	AC20
	AC16	AC16	AC16	AC17	AC17	AC17	AC18	AC18	AC18	AC19	AC19	AC19	AC20	AC20	AC20
	AC16	AC16	AC16	AC17	AC17	AC17	AC18	AC18	AC18	AC19	AC19	AC19	AC20	AC20	AC20
R2	AC7	AC7	AC7	AC14	AC14	AC14	AC1	AC1	AC1	AC6	AC6	AC6	AC13	AC13	AC13
	AC7	AC7	AC7	AC14	AC14	AC14	AC1	AC1	AC1	AC6	AC6	AC6	AC13	AC13	AC13
	AC7	AC7	AC7	AC14	AC14	AC14	AC1	AC1	AC1	AC6	AC6	AC6	AC13	AC13	AC13
	AC2	AC2	AC2	AC12	AC12	AC12	AC16	AC16	AC16	AC10	AC10	AC10	AC18	AC18	AC18
	AC2	AC2	AC2	AC12	AC12	AC12	AC16	AC16	AC16	AC10	AC10	AC10	AC18	AC18	AC18
	AC2	AC2	AC2	AC12	AC12	AC12	AC16	AC16	AC16	AC10	AC10	AC10	AC18	AC18	AC18
	AC11	AC11	AC11	AC9	AC9	AC9	AC4	AC4	AC4	AC20	AC20	AC20	AC3	AC3	AC3
	AC11	AC11	AC11	AC9	AC9	AC9	AC4	AC4	AC4	AC20	AC20	AC20	AC3	AC3	AC3
	AC11	AC11	AC11	AC9	AC9	AC9	AC4	AC4	AC4	AC20	AC20	AC20	AC3	AC3	AC3
	AC5	AC5	AC5	AC19	AC19	AC19	AC17	AC17	AC17	AC8	AC8	AC8	AC15	AC15	AC15
	AC5	AC5	AC5	AC19	AC19	AC19	AC17	AC17	AC17	AC8	AC8	AC8	AC15	AC15	AC15
	AC5	AC5	AC5	AC19	AC19	AC19	AC17	AC17	AC17	AC8	AC8	AC8	AC15	AC15	AC15

	AC9	AC9	AC9	AC1	AC1	AC1	AC15	AC15	AC15	AC10	AC10	AC10	AC3	AC3	AC3
	AC9	AC9	AC9	AC1	AC1	AC1	AC15	AC15	AC15	AC10	AC10	AC10	AC3	AC3	AC3
	AC9	AC9	AC9	AC1	AC1	AC1	AC15	AC15	AC15	AC10	AC10	AC10	AC3	AC3	AC3
	AC17	AC17	AC17	AC5	AC5	AC5	AC12	AC12	AC12	[illegible]	[illegible]	[illegible]	AC18	AC18	AC18
	AC17	AC17	AC17	AC5	AC5	AC5	AC12	AC12	AC12	[illegible]	[illegible]	[illegible]	AC18	AC18	AC18
R3	AC17	AC17	AC17	AC5	AC5	AC5	AC12	AC12	AC12	[illegible]	[illegible]	[illegible]	AC18	AC18	AC18
	AC14	AC14	AC14	AC8	AC8	AC8	AC19	AC19	AC19	AC13	AC13	AC13	AC11	AC11	AC11
	AC14	AC14	AC14	AC8	AC8	AC8	AC19	AC19	AC19	AC13	AC13	AC13	AC11	AC11	AC11
	AC14	AC14	AC14	AC8	AC8	AC8	AC19	AC19	AC19	AC13	AC13	AC13	AC11	AC11	AC11
	AC2	AC2	AC2	AC20	AC20	AC20	AC4	AC4	AC4	AC16	AC16	AC16	AC7	AC7	AC7
	AC2	AC2	AC2	AC20	AC20	AC20	AC4	AC4	AC4	AC16	AC16	AC16	AC7	AC7	AC7
	AC2	AC2	AC2	AC20	AC20	AC20	AC4	AC4	AC4	AC16	AC16	AC16	AC7	AC7	AC7

b. Pedigree of the Kadam genetic resources

Location	Seed Source	State	Latitude	Longitude	Altitude (m)
Tripura	FCRIAC1	Tripura	23°49′	91°17′	78
Khowai	FCRIAC2	Tripura	24°04′	91°36′	46
Tripura	FCRIAC3	Tripura	23°55′	91°50′	89
Mettupalayam	FCRIAC4	Tamil Nadu	11°17′	76°56′	330
Assam	FCRIAC5	Assam	26°33′	90°26′	69
Assam	FCRIAC6	Assam	26°33′	90°26′	69
Coimbatore	FCRIAC7	Tamil Nadu	11°17′	76°56′	330
Coimbatore	FCRIAC8	Tamil Nadu	11°17′	76°56′	330
Coimbatore	FCRIAC9	Tamil Nadu	11°17′	76°56′	330
Coimbatore	FCRIAC10	Tamil Nadu	11°17′	76°56′	330
Dapoli	FCRIAC11	Maharashtra	17°45′	73°11′	175
Maharashtra	FCRIAC12	Maharashtra	21°08′	79°05′	313
Tirupathi	FCRIAC13	Andhra Pradesh	13°37′	79°25′	155
Andhra Pradesh	FCRIAC14	Andhra Pradesh	13°37′	79°25′	155
Hyderabad	FCRIAC15	Andhra Pradesh	17°22′	78°29′	502
Andhra Pradesh	FCRIAC16	Andhra Pradesh	17°22′	78°29′	502
Karnal	FCRIAC17	Haryana	29°41′	76°59′	251
Haryana	FCRIAC18	Haryana	29°41′	76°59′	251
Thrissur	FCRIAC19	Kerala	10°31′	79°13′	54
Bangalore	FCRIAC20	Karnataka	12°58′	77°35′	910

c. Growth Evaluation of *Neolamarckia cadamba* seed sources at 5 years after planting

Name of the Seed source	Height (m)	Girth at Breast Height (cm)	Volume (m^3)	Biomass (kg/tree)
FCRIAC 1	6.22	38.56*	0.0520	103.95
FCRIAC 2	6.34	36.01	0.0466	93.15
FCRIAC 3	6.76	35.47	0.0476	95.28
FCRIAC 4	6.39	35.58	0.0455	90.95
FCRIAC 5	6.24	37.95*	0.0509	101.85
FCRIAC 6	6.44	34.99	0.0443	88.62
FCRIAC 7	7.53*	42.03*	0.0745*	149.06*
FCRIAC 8	5.72	31.22	0.0320	64.07
FCRIAC 9	5.99	34.98	0.0413	82.66
FCRIAC 10	5.86	32.52	0.0350	70.07
FCRIAC 11	6.19	35.99	0.0450	89.93
FCRIAC 12	6.66	36.74	0.0506	101.22
FCRIAC 13	7.12*	37.32*	0.0557	111.38
FCRIAC 14	6.15	41.27	0.0592*	118.47*
FCRIAC 15	6.04	32.28	0.0354	70.80
FCRIAC 16	7.24*	38.55*	0.0604*	120.74*
FCRIAC 17	6.42	41.86*	0.0631*	126.17*
FCRIAC 18	6.60	39.21*	0.0577*	115.49*
FCRIAC 19	6.43	37.69*	0.0515	102.96
FCRIAC 20	4.76	28.32	0.0215	43.01
Mean	**6.36**	**36.43**	**0.0485**	**96.99**
SEd	**0.33**	**0.28**	**0.0037**	**7.39**
CD (0.05)	**0.66**	**0.58**	**0.0075**	**14.95**

Inference / Findings

- The seed sources evaluation test conducted over a period of 5 years indicated significant variability among growth attributes for different biometric attributes.
- The height differed significantly due to seed sources and it ranged between 7.24 m and 4.76 m. Compared to the average

height of the seed sources (6.36 m), three seed sources viz., FCRIAC7 (7.53 m) followed by FCRIAC 16 (7.24 m) and FCRIAC 13 (7.12 m) registered significantly increased tree height.

- The seed sources exhibited great variability for girth at breast height. Among 20 seed sources evaluated, eight seed sources viz., FCRIAC 7 (42.03 cm), FCRIAC 17 (41.86 cm), FCRIAC 18 (39.21 cm), FCRIAC 1 (38.56 cm) FCRIAC 16 (38.55 cm) FCRIAC 5 (37.95 cm), FCRIAC 19 (37.69 cm) and FCRIAC 13 (37.32 cm) registered significantly higher GBH compared to the average GBH (36.43)
- Considering overall biomass into account, the seed sources exhibited wide variations which ranged between 149.06 kg/tree (FCRIAC 7) and 43.01 kg/tree (FCRIAC 20). Among 20 seed sources evaluated, the following 5 seed sources viz., FCRIAC 7 (149.06 kg) FCRIAC 17 (126.17 kg) FCRIAC16 (120.74 kg) FCRIAC14 (118.47 kg) and FCRIAC 18 (115.49 kg) recorded significantly higher biomass compared to the average biomass of the seed sources (96.99 kg / tree).
- Among the 5 superior seed sources, one seed source viz., FCRIAC 7 reigned supreme because the population from this seed source registered higher values for all biometric attributes investigated.
- Hence the Kadam seed source evaluation test identified the superiority of one seed source viz., FCRIAC 7 which is recommended for adoption in operation plantation and agroforestry development program.

III. PROGENY TEST (PT)

It is a test to verify the genetic worthiness of a parent through the performance of their offsprings or progenies.

Objectives of the progeny test

- To test the genetic worthiness of a parent
- To identify the best and suitable progenies for varied objectives
- To establish seedling seed orchard
- To establish seed production area
- To conserve the genetic resources of the identified species.

Seed collection for Progeny test

- For any progeny test, selection of superior genetic stock (Plus Trees) is very important. In the identified species, the plus trees can be selected either from natural forest or from plantations using regression and comparison tree method respectively for these two forest types.
- The selection of tree is based on breeding objective envisaged. However the most selection programme is based on superiority of the selected individual's interms of height, GBH, Tree form, free form pest and diseases etc., compared to the neighbourers.
- For breeding towards wood quality, the selection should be based on wood quality characterization of the identified trees.
- These selected individual trees are termed as plus trees.
- From these plus trees seeds are collected and labled as progenies.

Plot size

- Similar to other tests, the plot size also variable and depending on the number of progenies it may be a single tree based or row column or rectangular or square plot based sizes.
- If the numbers of the progenies are more i.e. more than 100, then it is better to have single tree or row column based plot size.
- If the number of progenies is less. (<50) it is better to have square or rectangular plot.

Design

- The simple and most appropriate design is Randomized complete block design.

Method of progeny test

Number of progenies	–	15 to any number
Design	–	RBD (Single tree plot)
Method	–	• Single tree plot (if number of progenies are more) • Line plot (if area is restricted) • Block plot (Ideal method)

Spacing	–	3 m × 3m but variable due to species and objective of breeding
Period	–	0.5 to 1 × rotation age

PRACTICAL FIELD TEST OF REDSANDER PROGENY TESTS

Breeding Objective	–	To screen HYSR progeny
Name of the Species	–	Redsanders
Name of the Evaluation	-	Progeny Test
Location	–	Forest College and Research Institute, Mettupalayam
	–	Latitude: 11°18' N and Longitude: 76°59' E
Number of progenies	–	24
Design	–	RBD
Method	–	Block plot (9 tree plot at 3 x 3)
Spacing	–	3 m × 3m
Period	–	15 years

Growth attributes evaluated

- Tree height (m)
- Clear bole height (m)
- GBH (cm)
- Basal girth (cm)
- Number of branches
- Clear bole volume (m^3)
- Volume (m^3)

a. Layout of the Redsanders Progeny Test (PT)

	RS21	RS21	RS21	RS01	RS01	RS01	RS24	RS24	RS24	RS02	RS02	RS02	RS09	RS09	RS09	RS19	RS19	RS19
	RS21	RS21	RS21	RS01	RS01	RS01	RS24	RS24	RS24	RS02	RS02	RS02	RS09	RS09	RS09	RS19	RS19	RS19
	RS21	RS21	RS21	RS01	RS01	RS01	RS24	RS24	RS24	RS02	RS02	RS02	RS09	RS09	RS09	RS19	RS19	RS19
	RS06	RS06	RS06	RS17	RS17	RS17	RS08	RS08	RS08	RS16	RS16	RS16	RS11	RS11	RS11	RS05	RS05	RS05
	RS06	RS06	RS06	RS17	RS17	RS17	RS08	RS08	RS08	RS16	RS16	RS16	RS11	RS11	RS11	RS05	RS05	RS05
R1	RS06	RS06	RS06	RS17	RS17	RS17	RS08	RS08	RS08	RS16	RS16	RS16	RS11	RS11	RS11	RS05	RS05	RS05
	RS12	RS12	RS12	RS10	RS10	RS10	RS03	RS03	RS03	RS13	RS13	RS13	RS07	RS07	RS07	RS22	RS22	RS22
	RS12	RS12	RS12	RS10	RS10	RS10	RS03	RS03	RS03	RS13	RS13	RS13	RS07	RS07	RS07	RS22	RS22	RS22
	RS12	RS12	RS12	RS10	RS10	RS10	RS03	RS03	RS03	RS13	RS13	RS13	RS07	RS07	RS07	RS22	RS22	RS22
	RS04	RS04	RS04	RS23	RS23	RS23	RS15	RS15	RS15	RS20	RS20	RS20	RS18	RS18	RS18	RS14	RS14	RS14
	RS04	RS04	RS04	RS23	RS23	RS23	RS15	RS15	RS15	RS20	RS20	RS20	RS18	RS18	RS18	RS14	RS14	RS14
	RS04	RS04	RS04	RS23	RS23	RS23	RS15	RS15	RS15	RS20	RS20	RS20	RS18	RS18	RS18	RS14	RS14	RS14
	RS02	RS02	RS02	RS13	RS13	RS13	RS22	RS22	RS22	RS05	RS05	RS05	RS16	RS16	RS16	RS21	RS21	RS21
	RS02	RS02	RS02	RS13	RS13	RS13	RS22	RS22	RS22	RS05	RS05	RS05	RS16	RS16	RS16	RS21	RS21	RS21
	RS02	RS02	RS02	RS13	RS13	RS13	RS22	RS22	RS22	RS05	RS05	RS05	RS16	RS16	RS16	RS21	RS21	RS21
	RS08	RS08	RS08	RS24	RS24	RS24	RS10	RS10	RS10	RS19	RS19	RS19	RS12	RS12	RS12	RS07	RS07	RS07
	RS08	RS08	RS08	RS24	RS24	RS24	RS10	RS10	RS10	RS19	RS19	RS19	RS12	RS12	RS12	RS07	RS07	RS07
R2	RS08	RS08	RS08	RS24	RS24	RS24	RS10	RS10	RS10	RS19	RS19	RS19	RS12	RS12	RS12	RS07	RS07	RS07
	RS11	RS11	RS11	RS15	RS15	RS15	RS06	RS06	RS06	RS14	RS14	RS14	RS23	RS23	RS23	RS03	RS03	RS03
	RS11	RS11	RS11	RS15	RS15	RS15	RS06	RS06	RS06	RS14	RS14	RS14	RS23	RS23	RS23	RS03	RS03	RS03
	RS11	RS11	RS11	RS15	RS15	RS15	RS06	RS06	RS06	RS14	RS14	RS14	RS23	RS23	RS23	RS03	RS03	RS03
	RS20	RS20	RS20	RS04	RS04	RS04	RS01	RS01	RS01	RS17	RS17	RS17	RS09	RS09	RS09	RS18	RS18	RS18
	RS20	RS20	RS20	RS04	RS04	RS04	RS01	RS01	RS01	RS17	RS17	RS17	RS09	RS09	RS09	RS18	RS18	RS18
	RS20	RS20	RS20	RS04	RS04	RS04	RS01	RS01	RS01	RS17	RS17	RS17	RS09	RS09	RS09	RS18	RS18	RS18

R3	RS06	RS06	RS06	RS19	RS19	RS19	RS03	RS03	RS03	RS07	RS07	RS07	RS02	RS02	RS02	RS23	RS23	RS23
	RS06	RS06	RS06	RS19	RS19	RS19	RS03	RS03	RS03	RS07	RS07	RS07	RS02	RS02	RS02	RS23	RS23	RS23
	RS06	RS06	RS06	RS19	RS19	RS19	RS03	RS03	RS03	RS07	RS07	RS07	RS02	RS02	RS02	RS23	RS23	RS23
	RS21	RS21	RS21	RS11	RS11	RS11	RS16	RS16	RS16	RS12	RS12	RS12	RS20	RS20	RS20	RS10	RS10	RS10
	RS21	RS21	RS21	RS11	RS11	RS11	RS16	RS16	RS16	RS12	RS12	RS12	RS20	RS20	RS20	RS10	RS10	RS10
	RS21	RS21	RS21	RS11	RS11	RS11	RS16	RS16	RS16	RS12	RS12	RS12	RS20	RS20	RS20	RS10	RS10	RS10
	RS17	RS17	RS17	RS01	RS01	RS01	RS09	RS09	RS09	RS05	RS05	RS05	RS08	RS08	RS08	RS15	RS15	RS15
	RS17	RS17	RS17	RS01	RS01	RS01	RS09	RS09	RS09	RS05	RS05	RS05	RS08	RS08	RS08	RS15	RS15	RS15
	RS17	RS17	RS17	RS01	RS01	RS01	RS09	RS09	RS09	RS05	RS05	RS05	RS08	RS08	RS08	RS15	RS15	RS15
	RS24	RS24	RS24	RS04	RS04	RS04	RS14	RS14	RS14	RS18	RS18	RS18	RS13	RS13	RS13	RS22	RS22	RS22
	RS24	RS24	RS24	RS04	RS04	RS04	RS14	RS14	RS14	RS18	RS18	RS18	RS13	RS13	RS13	RS22	RS22	RS22
	RS24	RS24	RS24	RS04	RS04	RS04	RS14	RS14	RS14	RS18	RS18	RS18	RS13	RS13	RS13	RS22	RS22	RS22

b. Pedigree of the Redsanders progenies

S. No	Progeny	Source	District	Latitude	Longitude
1.	TNRS -1	Ammur I	Vellore	12° 46' 09.78'' N	78° 44' 20.69'' E
2.	TNRS -2	Ammur I	Vellore	12° 49' 04.07'' N	78° 47' 54.72'' E
3.	TNRS -3	Ammur I	Vellore	12° 46' 12.57'' N	78° 51' 18.60'' E
4.	TNRS -4	Ammur I	Vellore	12° 44' 31.19'' N	78° 52' 24.06'' E
5.	TNRS -5	Ammur I	Vellore	12° 45' 45.20'' N	78° 51' 12.36'' E
6.	TNRS -6	Ammur I	Vellore	12° 44' 10.90'' N	78° 52' 57.31'' E
7.	TNRS -7	Ammur I	Vellore	12° 48' 36.05'' N	78° 49' 18.05'' E
8.	TNRS -8	Ammur II	Vellore	12° 42' 33.74'' N	78° 46' 53.59'' E
9.	TNRS -9	Ammur II	Vellore	12° 39' 21.00'' N	78° 50' 48.40'' E
10.	TNRS -10	Pitchanur	Vellore	12° 58' 00.82'' N	78° 52' 16.31'' E
11.	TNRS -11	Pitchanur	Vellore	12° 58' 08.65'' N	78° 52' 54.63'' E
12.	TNRS -12	Pitchanur	Vellore	12° 57' 36.72'' N	78° 53' 08.84'' E
13.	TNRS -13	Pitchanur	Vellore	12° 57' 56.60'' N	78° 52' 43.50'' E
14.	TNRS -14	Banavaram	Vellore	13° 01' 18.95'' N	79° 47' 50.31'' E
15.	TNRS -15	Banavaram	Vellore	13° 01' 11.36'' N	79° 30' 08.50'' E
16.	TNRS -16	Banavaram	Vellore	13° 01' 31.48'' N	79° 29' 26.93'' E
17.	TNRS -17	Veerichettipalli	Vellore	13° 03' 19.23'' N	78° 53' 02.01'' E
18.	TNRS -18	Veerichettipalli	Vellore	13° 04' 18.66'' N	78° 52' 55.38'' E
19.	TNRS -19	Veerichettipalli	Vellore	13° 04' 32.25'' N	78° 51' 49.21'' E

S. No	Progeny	Source	District	Latitude	Longitude
20.	TNRS -20	Kalapadi	Vellore	13° 01' 22.92'' N	78° 49' 41.04'' E
21.	TNRS -21	Kalapadi	Vellore	13° 02' 12.56'' N	78° 47' 27.79'' E
22.	TNRS -22	Kalapadi	Vellore	13° 02' 56.06'' N	78° 47' 09.41'' E
23.	TNRS -23	Kalapudur	Vellore	12° 59' 19.77'' N	79° 12' 27.21'' E
24.	TNRS -24	Kalapudur	Vellore	13° 00' 41.07'' N	79° 13' 00.66'' E

c. Progeny Variation for growth attributes in Redsanders at 15 Years after planting

S. No	Genotypes	Tree height (m)	Clear bole height (m)	GBH (cm)	Clear bole volume (m^3)	Volume (m^3)
1.	TNRS -1	11.88*	6.6*	74.8*	0.291*	0.533*
2.	TNRS -2	7.92*	4.4*	53.6	0.103	0.190
3.	TNRS -3	5.94	3.3	37.2	0.038	0.071
4.	TNRS -4	6.93	3.8	34.4	0.037	0.066
5.	TNRS -5	7.43	4.1	40.6	0.060	0.107
6.	TNRS -6	5.94	3.3	34.8	0.033	0.058
7.	TNRS -7	7.92*	4.4*	47.7	0.084	0.149
8.	TNRS -8	5.94	3.3	31.0	0.026	0.046
9.	TNRS -9	8.91*	4.9*	73.6*	0.215*	0.386*
10.	TNRS -10	7.92*	4.4*	61.0*	0.131*	0.237*
11.	TNRS -11	9.90*	5.5*	66.8*	0.194*	0.351*
12.	TNRS -12	8.91*	4.9*	53.2	0.112	0.202
13.	TNRS -13	5.94	3.3	26.3	0.018	0.034
14.	TNRS -14	7.92*	4.4*	35.1	0.043	0.078
15.	TNRS -15	6.93	3.8	36.2	0.041	0.073
16.	TNRS -16	7.92*	4.4*	54.0*	0.101	0.186
17.	TNRS -17	7.92*	4.4*	34.7	0.042	0.077
18.	TNRS -18	6.93	3.8	49.3	0.075	0.135
19.	TNRS -19	5.94	3.3	23.4	0.014	0.027
20.	TNRS -20	6.44	3.5	37.7	0.041	0.075
21.	TNRS -21	5.94	3.3	30.2	0.024	0.044
22.	TNRS -22	7.92*	4.4*	34.9	0.044	0.079
23.	TNRS -23	8.91*	4.9*	56.2*	0.124*	0.228*
24.	TNRS -24	9.90*	5.5*	56.7*	0.139*	0.255*
	Population Mean	**7.67**	**4.2**	**45.2**	**0.084**	**0.153**
	SEd	**0.06**	**0.06**	**4.3**	**0.016**	**0.309**
	CD (0.5)	**0.13**	**0.10**	**8.5**	**0.032**	**0.061**

Inference / Findings

- The redsanders progeny test was carried out for fifteen years and the study indicated that the progenies exhibited significant variations for all the biometric attributes investigated. The progenies were compared with population mean for their significance as there is no standard variety available for comparison.
- The plant height of redsanders progenies exhibited significant variation among them and the height of the progenies ranged between 11.88 m and 5.94 m. Compared to the average height of the progenies (7.67 m), 13 progenies viz., TNRS-1 (11.88 m), TNRS-24 (9.90 m), TNRS-11 (9.90 m), TNRS-12(8.91 m), TNRS-23(8.91 m), TNRS-9 (8.91 m), TNRS-22(7.92 m), TNRS-10(7.92 m), TNRS-7(7.92 m), TNRS-2(7.92 m), TNRS-16(7.92 m), TNRS-17 (7.92 m) and TNRS-14 (7.90 m) registered significantly higher tree height compared to the progeny mean height (7.67 m).
- Similarly, the redsander progenies exhibited wide variations for girth at breast height and the progenies recorded an average girth of 45.2 cm after 15 years of evaluation. Compared to the average girth, 7 progenies viz, TNRS-1 (74.8 cm), TNRS-9 (73.6 cm), TNRS-11 (66.8 cm), TNRS-10 (61.0 cm), TNRS-24 (56.7 cm), TNRS-23 (56.2 cm) and TNRS-16 (54.0 cm) registered significantly higher girth at breast height.
- Similar trend was also observed in clear bole volume wherein 6 progenies viz., TNRS-1 (0.291 m^3), TNRS-9 (0.215 m^3) TNRS-11(0.194 m^3) TNRS-24 (0.139 m^3) TNRS-10 (0.131 m^3) and TNRS-23 (0.124 m^3) exhibited significantly higher clear bole volume.
- The volume differed significantly due to progenies and the average volume registered by the progenies is 0.153 m^3 / tree. Compared to the average volume of progenies, 6 progenies viz, TNRS-1 (0.533 m^3), TNRS-9 (0.386 m^3), TNRS-11 (0.351 m^3), TNRS-24 (0.255 m^3), TNRS-10 (0.237 m^3) and TNRS-23 (0.228 m^3) registered significantly higher volume.
- Considering all the biometric attributes into account, the redsanders progeny evaluation test identified the superiority of TNSR1 interms of all growth attributes and this progeny is

recommended for adoption in operational plantation development program.

IV. CLONAL TEST (CT)

In the recent past, the clonal test for most commercial tree species has gained potential attraction to achieve productivity improvement and also to evolve clones suitable for biotic and abiotic resistance and tolerance.

Selection and deployment of clones

- The most important activity of the clonal test is inventory and selection of potential clones.
- The clones can be identified based on individual superiority compared to their corresponding neighborers.
- The tree can be selected from provenance trial or seed sources or progenies or hybrid progenies and deployed as a clone through clonal multiplication.
- Most hybrid vigour or heterosis is exploited through clonal technology.
- Once trees are indentified, the vegetative parts can be collected and multiplied through micro propagation or tissue culture which may act as a clonal plantlets for that particular individual tree or clone.
- These plants can be planted in mother garden and mass multiplied using mini clonal technology
- Alternatively the identified trees are felled at 15 cm above ground level and coppice shoots are induced.
- These coppice shoots are rooted and the rooted plantlets are again planted in the mother garden and multiplication process continued through mini clonal technology.
- A minimum of 15-30 clones are required for a standard clonal test. Higher the number of clones, more will be the genetic gain.

Plot size

- For any clonal test, the plot size is variable like single tree or row coloumn or rectangular or square type.

- If the numbers of clones are between 15 and 30, square plot size is preferred.
- If the number of clones is beyond 100, then either row column or single tree based plot size is enough

Design

- The most preferred design for any clonal evaluation test is randomized complete block design in which the experimental area is divided into equisized blocks (Replications).
- Each block is further divided into number of equal sized blocks depending on the number of clones deployed for evaluation.

Standard Clonal Test

Number of clones	–	15 to 30 (Minimum)
Number of replication	–	3 to 10
Design	–	Randomized Block Design
Spacing	–	1.5 m × 1.5 m for SR species 3 m × 3 m for MR species 5 m × 5 m for LR species
Plot size	–	Square Plot
Period of evaluation	–	SR & MR Species : Full rotation period LR Species : Half rotation period

Evaluation of Characters / Traits

BIOMETRIC EVALUATION

- Tree height
- Clean bole height
- Girth at breast height
- Basal girth
- Number of branches
- Volume
- Clean bole volume
- Yield

WOOD QUALITY EVALUATION

i. **Physical Properties**

- Moisture content
- Basic Density
- Bulk Density

ii. **Chemical Properties**

- Ash content
- Acid insoluble lignin
- Holocellulose
- Pentasans

iii. **Anatomical Properties**

- Fibre length
- Fibre diameter
- Fibre lumen width
- Fibre wall thickness
- Vessel length
- Vessel diameter
- Ray morphology

iv. **Mechanical Properties**

- Static bending Strength
- Compression strength parallel to grain
- Compression strength perpendicular to grain
- Tension strength parallel to grain
- Tension strength perpendicular to grain
- MOE parallel to grain
- MOE perpendicular to grain
- MOR parallel to grain
- MOR perpendicular to grain

COMBINED EVALUATION

- Productivity and Wood Quality Analysis

PRACTICAL FIELD TEST OF CASUARINA CLONAL TEST (CT)

Breeding Objective	–	To screen HYSR clone
Name of the Species	–	Casuarina
Name of the Evaluation	–	Clonal Test
Location	–	Forest College and Research Institute, Mettupalayam
	–	Latitude: 11°18' N and Longitude: 76°59' E
Number of progenies	–	21
Design	–	RBD
Method	–	Block plot (9 tree plot at 3 × 3)
Number of Replication	–	3
Spacing	–	3 m × 3m
Period	–	7 years

Growth attributes evaluated

- Tree height (m)
- Clear bole height (m)
- GBH (cm)
- Basal girth (cm)
- Number of branches
- Clear bole volume (m^3)
- Volume (m^3)

a. Layout of the clonal evaluation test

ROAD

NURSERY														
CJ3	CJ3	CJ3	CJ16	CJ16	CJ16	CJ13	CJ13	CJ13	CJ3	CJ3	CJ3	CJ6	CJ6	CJ6
CJ3	CJ3	CJ3	CJ16	CJ16	CJ16	CJ13	CJ13	CJ13	CJ3	CJ3	CJ3	CJ6	CJ6	CJ6
CJ3	CJ3	CJ3	CJ16	CJ16	CJ16	CJ13	CJ13	CJ13	CJ3	CJ3	CJ3	CJ6	CJ6	CJ6
CJ6	CJ6	CJ6	CJ11	CJ11	CJ11	CJ24	CJ24	CJ24	CJ27	CJ27	CJ27	CJ14	CJ14	CJ14
CJ6	CJ6	CJ6	CJ11	CJ11	CJ11	CJ24	CJ24	CJ24	CJ27	CJ27	CJ27	CJ14	CJ14	CJ14
CJ6	CJ6	CJ6	CJ11	CJ11	CJ11	CJ24	CJ24	CJ24	CJ27	CJ27	CJ27	CJ14	CJ14	CJ14
CJ8	CJ8	CJ8	CJ12	CJ12	CJ12	CJ26	CJ26	CJ26	CJSS	CJSS	CJSS	CJ8	CJ8	CJ8
CJ8	CJ8	CJ8	CJ12	CJ12	CJ12	CJ26	CJ26	CJ26	CJSS	CJSS	CJSS	CJ8	CJ8	CJ8
CJ8	CJ8	CJ8	CJ12	CJ12	CJ12	CJ26	CJ26	CJ26	CJSS	CJSS	CJSS	CJ8	CJ8	CJ8
CJ9	CJ9	CJ9	CJ14	CJ14	CJ14	CJ18	CJ18	CJ18	CJ30	CJ30	CJ30	CJ24	CJ24	CJ24
CJ9	CJ9	CJ9	CJ14	CJ14	CJ14	CJ18	CJ18	CJ18	CJ30	CJ30	CJ30	CJ24	CJ24	CJ24
CJ9	CJ9	CJ9	CJ14	CJ14	CJ14	CJ18	CJ18	CJ18	CJ30	CJ30	CJ30	CJ24	CJ24	CJ24
CJ10	CJ10	CJ10	Control	Control	Control	CJ3	CJ3	CJ3	CJ10	CJ10	CJ10	CJ11	CJ11	CJ11
CJ10	CJ10	CJ10	Control	Control	Control	CJ3	CJ3	CJ3	CJ10	CJ10	CJ10	CJ11	CJ11	CJ11
CJ10	CJ10	CJ10	Control	Control	Control	CJ3	CJ3	CJ3	CJ10	CJ10	CJ10	CJ11	CJ11	CJ11
CJ11	CJ11	CJ11	CESS	CESS	CESS	CJ9	CJ9	CJ9	CJ18	CJ18	CJ18	CJ9	CJ9	CJ9
CJ11	CJ11	CJ11	CESS	CESS	CESS	CJ9	CJ9	CJ9	CJ18	CJ18	CJ18	CJ9	CJ9	CJ9
CJ11	CJ11	CJ11	CESS	CESS	CESS	CJ9	CJ9	CJ9	CJ18	CJ18	CJ18	CJ9	CJ9	CJ9
CJ12	CJ12	CJ12	CJSS	CJSS	CJSS	CJ8	CJ8	CJ8	CJ26	CJ26	CJ26	CESS	CESS	CESS
CJ12	CJ12	CJ12	CJSS	CJSS	CJSS	CJ8	CJ8	CJ8	CJ26	CJ26	CJ26	CESS	CESS	CESS
CJ12	CJ12	CJ12	CJSS	CJSS	CJSS	CJ8	CJ8	CJ8	CJ26	CJ26	CJ26	CESS	CESS	CESS
CJ13	CJ13	CJ13	CJ30	CJ30	CJ30	CJ10	CJ10	CJ10	CJ12	CJ12	CJ12	CJ28	CJ28	CJ28
CJ13	CJ13	CJ13	CJ30	CJ30	CJ30	CJ10	CJ10	CJ10	CJ12	CJ12	CJ12	CJ28	CJ28	CJ28

CJ13	CJ13	CJ13	CJ30	CJ30	CJ30	CJ10	CJ10	CJ10	CJ12	CJ12	CJ12	CJ28	CJ28	CJ28
CJ14	CJ14	CJ14	CJ29	CJ29	CJ29	CJ6	CJ6	CJ6	CJ28	CJ28	CJ28	CJ13	CJ13	CJ13
CJ14	CJ14	CJ14	CJ29	CJ29	CJ29	CJ6	CJ6	CJ6	CJ28	CJ28	CJ28	CJ13	CJ13	CJ13
CJ14	CJ14	CJ14	CJ29	CJ29	CJ29	CJ6	CJ6	CJ6	CJ28	CJ28	CJ28	CJ13	CJ13	CJ13
CJ16	CJ16	CJ16	CJ28	CJ28	CJ28	CJ27	CJ27	CJ27	Control	Control	Control	Control	Control	Control
CJ16	CJ16	CJ16	CJ28	CJ28	CJ28	CJ27	CJ27	CJ27	Control	Control	Control	Control	Control	Control
CJ16	CJ16	CJ16	CJ28	CJ28	CJ28	CJ27	CJ27	CJ27	Control	Control	Control	Control	Control	Control
CJ18	CJ18	CJ18	CJ27	CJ27	CJ27	CJ29	CJ29	CJ29	CESS	CESS	CESS	CJ29	CJ29	CJ29
CJ18	CJ18	CJ18	CJ27	CJ27	CJ27	CJ29	CJ29	CJ29	CESS	CESS	CESS	CJ29	CJ29	CJ29
CJ18	CJ18	CJ18	CJ27	CJ27	CJ27	CJ29	CJ29	CJ29	CESS	CESS	CESS	CJ29	CJ29	CJ29
CJ24	CJ24	CJ24	CJ26	CJ26	CJ26	CJ30	CJ30	CJ30	CJSS	CJSS	CJSS	CJ16	CJ16	CJ16
CJ24	CJ24	CJ24	CJ26	CJ26	CJ26	CJ30	CJ30	CJ30	CJSS	CJSS	CJSS	CJ16	CJ16	CJ16
CJ24	CJ24	CJ24	CJ26	CJ26	CJ26	CJ30	CJ30	CJ30	CJSS	CJSS	CJSS	CJ16	CJ16	CJ16

SISSOO FIELD

b. Growth evaluation of various clones of casuarina

S. No	Clones	Height (m)	Clean Bole Height (m)	Girth at Breast Height (cm)	Volume (m^3)
1.	CJ3	10.43*	5.17	35.77*	0.0830*
2.	CJ6	12.42*	7.71*	30.17*	0.0787*
3.	CJ8	11.89*	7.28*	39.19*	0.1201*
4.	CJ9	10.01*	5.65*	27.88*	0.0486*
5.	CJ10	8.33	6.35*	31.01*	0.0523*
6.	CJ11	9.30	6.19*	34.72*	0.0761*
7.	CJ12	10.85*	7.25*	31.55*	0.0748*
8.	CJ13	9.63*	5.53*	30.28*	0.0611*
9.	CJ14	10.27*	6.87*	33.35*	0.0781*
10.	CJ16	13.13*	8.45*	44.90*	0.1866*
11.	CJ18	9.52*	6.08*	31.63*	0.0695*
12.	CJ24	10.63*	6.70*	40.13*	0.1050*
13.	CJ26	8.92	6.17*	31.08*	0.0520*
14.	CJ27	10.50*	6.56*	34.17*	0.0903*
15.	CJ28	8.90	5.57*	31.53*	0.0511*
16.	CJ29	9.47*	5.78*	29.31*	0.0592*
17.	CJ30	7.22	4.81	27.44*	0.0357
18.	CJ33	7.33	4.17	27.00	0.0308
19.	CJSS	10.50*	7.17*	37.67*	0.0902*
20.	CESS	10.45*	5.82*	33.03*	0.0698*
21.	Control	8.61	4.83	24.56	0.0307
	Mean	**9.92**	**6.19**	**32.68**	**0.0735**
	SE.d	**0.39**	**0.27**	**1.38**	**0.0040**
	CD (0.05)	**0.79**	**0.54**	**2.80**	**0.0081**

Inference / Findings

- Twenty Casuarina clones were evaluated in a standard clonal test incorporating a check variety for comparison. The clone exhibited significantly higher values compared to the control variety and indicates that all the tested clones are superior.

- The clones differed significantly for height which ranged between 13.13 m and 7.22 m. Compared to control variety 14 clones exhibited significantly higher tree height.
- Similarly the clones differed significantly for clear bole height and 17 clones exhibited significantly higher clear bole height.
- The clones varied significantly due to Girth at breast height and out of 20 clones evaluated almost all the clones reigned supreme compared to the control variety except CJ 33.
- In volume also except 2 clones viz., CJ30 and CJ33 all other 18 clones exhibited significantly higher volume.
- In a holistic perspective almost all the clones barring CJ30 and CJ33 are amenable for deployment in operational clonal plantation developmental activities.

SUMMARY

The growing demand for wood and wood products and the restrictions on felling of natural forests coupled with changes in policy and legal issues have compelled the wood based industries and other stake holders to implement plantation forestry and various agroforestry systems. This intensive plantation forestry programme requires availability of superior planting stocks. Till the recent past, most plantation activities in the country have been established using unimproved seed sources thus resulting in poor productivity and detracted the attraction of plantation and agroforestry promotion. Under such circumstances, several tree breeding and improvement programme were implemented within and outside the country deploying traditional method of selection, hybridization and biotechnological tools. These breeding methods have evolved high yielding short rotation varieties. These breeding methods have exploited natural variability leading to the development of high yielding varieties, genetic resources with improved wood quality and genetic resources resistance towards biotic and abiotic influences. Among several tree breeding programme, the potential and utility of provenances/ seed sources/ progenies/ clones are very well established in wide range of species of both conifers and tropical hardwoods. However the practical aspects and methods of several breeding programme are not available in the form of a text book for ready reference. Hence the current chapter has been designed and presented with the practical aspects of the common breeding tests practiced in forestry and demonstrated with actual field data collected from various species and tests respectively. This chapter will be of great use to the students of forestry programme, forestry researchers and wood based industries involved in varietal development programme and professional foresters as a source of great piece of information both for understanding the breeding process and apply them for research as well as field establishment.

7

STATISTICAL TOOLS AND TECHNIQUES APPLIED IN TREE IMPROVEMENT PROGRAMME

INTRODUCTION

Tree breeding and improvement programme has been carried out across the country in 25 prioritized species. The improvement programme has mandated development of varieties for productivity improvement coupled with improved wood quality. This besides, development of resistance varieties for both biotic and abiotic stresses also form a major scope of tree improvement programme. The evaluation and identification of high cellulose and low lignin varieties for varied industrial utility, low bark content, higher density, increased fuel value are the major factors of concern in most tree improvement programme. During the journey of short term and long term tree improvement programme, the tree breeders and professional foresters deal with voluminous biometric, physiological and quality data which demand systematic analysis to cater to the needs of scientific documentation and reporting. Hence there is a need to have basic statistical knowledge coupled with applied experimental design to act as a scientific tool for the tree breeders, professional foresters and the graduating scholars. Hence this chapter on "STATISTICAL TOOLS AND TECHNIQUES" has been conceived and presented with suitable examples related to handling of tree improvement data which may have profound utility.

BASIC STATISTICAL TOOLS

The knowledge on basics of statistical tool is a prerequisite for the tree breeders in order to learn the biometric and quantitative genetics of most tree crop evaluation programme. Accordingly, the following basic statistical tools are presented with suitable examples to benefit the readers and practicing tree breeders.

i. Arithmetic Mean (AM)
ii. Median
iii. Mode
iv. Standard Deviation (SD)
v. Variance
vi. Coefficient of Variation
vii. Critical Difference

i. ARITHMETIC MEAN OR MEAN

Arithmetic mean or simply the mean of a variable is defined as the sum of the observations divided by the number of observations.

$$Arithmetic\ Mean = \frac{\sum X}{n}$$

Where

X is the observations

n is the number of observations

Merits

- It is rigidly defined.
- It is easy to understand and easy to calculate.
- If the number of items is sufficiently large, it is more accurate and more reliable.
- It is a calculated value and is not based on its position in the series.
- It is possible to calculate even if some of the details of the data are lacking.
- Of all averages, it is affected least by fluctuations of sampling.
- It provides a good basis for comparison.

Demerits

- It cannot be obtained by inspection or located through a frequency graph.

- It cannot be in the study of qualitative phenomena not capable of numerical measurement i.e. Intelligence, beauty, honesty etc.,
- It can ignore any single item only at the risk of losing its accuracy.
- It is affected very much by extreme values.
- It cannot be calculated for open-end classes.
- It may lead to fallacious conclusions, if the details of the data from which it is computed are not given.

Example

The heights of the Melia Seed sources are given below table. From the below table, calculate the following

i. Arithmetic Mean (AM)

ii. Median

iii. Mode

iv. Standard Deviation (SD)

v. Variance

vi. Coefficient of Variation

vii. Critical Difference

Seed sources	**Tree height of Melia (m)**		
	R1	**R2**	**R3**
MD 1	14.72	17.33	13.50
MD 2	14.94	17.46	16.48
MD 3	15.33	13.78	15.39
MD 4	9.94	13.83	17.33
MD 5	11.73	18.11	15.50
MD 6	13.27	16.44	13.11
MD 7	11.17	14.50	14.22
MD 8	12.52	17.33	16.28
MD 9	14.09	12.06	15.67
MD 10	11.13	15.78	13.17
MD 11	13.76	16.18	16.53

Seed sources	Tree height of Melia (m)		
	R1	R2	R3
MD 12	13.80	12.64	14.30
MD 13	14.73	16.18	15.89
MD 14	15.39	18.50	14.22
MD 15	14.53	16.56	15.50
MD 16	14.71	16.61	12.36
MD 17	14.06	14.50	9.52
MD 18	12.83	16.94	16.07
MD 19	11.78	15.44	16.78
MD 20	13.17	14.56	13.33

Arithmetic Mean (AM)

$$\text{AM} = \frac{\sum x}{n}$$

$$= \frac{14.72 + 14.94 + 15.38 \ldots\ldots + 16.78 + 13.33}{60}$$

Mean = 14.63 m

ii. MEDIAN

The median is the middle most item that divides the group into two equal parts, one part comprising all values greater, and the other, all values less than that item.

- Arrange the given values in the ascending order.
- If the numbers of values are odd, median is the middle value
- If the numbers of values are even, median is the mean of middle two values.

By formula

When n is odd,

$$\text{Median} = \frac{n+1}{2} th\ value$$

When n is even,

$$\text{Median} = \frac{n}{2}\ and\ \frac{n}{2} + 1th\ value$$

Merits of Median

- Median is not influenced by extreme values because it is a positional average.
- Median can be calculated in case of distribution with open-end intervals.
- Median can be located even if the data are incomplete.

Demerits of Median

- A slight change in the series may bring drastic change in median value.
- In case of even number of items or continuous series, median is an estimated value other than any value in the series.
- It is not suitable for further mathematical treatment except its use in calculating mean deviation.
- It does not take into account all the observations.

Example

Total No. of Parameters = 60

$$= \frac{60}{2} = 30.5$$

Write in Ascending / Descending orders

Then select 30.5th number. So we select the average of 30th, 31st number

9.52, 9.94, 11.13, 11.17, 11.73, 11.78, 12.06, 12.36, 12.52, 12.83, 12.94, 13.11, 13.17, 13.17, 13.27, 13.33, 13.50, 13.76, 13.78, 13.80, 13.83, 14.06, 14.09, 14.22, 14.22, 14.30, 14.50, 14.50, 14.53, **14.56, 14.71**, 14.72, 14.73, 14.94, 15.33, 15.39, 15.39, 15.44, 15.50, 15.50, 15.67, 15.79, 15.89, 16.07, 16.18, 16.18, 16.28, 16.44, 16.48, 16.53, 16.56, 16.61, 16.78, 16.94, 17.33, 17.33, 17.33, 17.46, 18.11, 18.50

$$= \frac{14.56}{2}$$

Median = 14.64

iii. MODE

The mode refers to that value in a distribution, which occur most frequently. It is an actual value, which has the highest concentration of items in and around it. It shows the centre of concentration of the frequency in around a given value. Therefore, where the purpose is to know the point of the highest concentration it is preferred.

Example

Mode

9.52, 9.94, 11.13, 11.17, 11.73, 11.78, 12.06, 12.36, 12.52, 12.83, 12.94, 13.11, 13.17, 13.17, 13.27, 13.33, 13.50, 13.76, 13.78, 13.80, 13.83, 14.06, 14.09, 14.22, 14.22, 14.30, 14.50, 14.50, 14.53, 14.56, 14.71, 14.72, 14.73, 14.94, 15.33, 15.39, 15.39, 15.44, 15.50, 15.50, 15.67, 15.79, 15.89, 16.07, 16.18, 16.18, 16.28, 16.44, 16.48, 16.53, 16.56, 16.61, 16.78, 16.94, **17.33, 17.33, 17.33,** 17.46, 18.11, 18.50

Most frequent Number is 17.33 (Three times)

The mode is 17.33

iv. DISPERSION

The averages are representatives of a frequency distribution. But they fail to give a complete picture of the distribution.

The averages help the breeder in considering a mass of data into a single value which represent the entire population but fail to reveal the entire picture of distribution. The degree of the scatteredness or variance of the individual observations for their average is called dispersion.

Characteristics of a good measure of dispersion

An ideal measure of dispersion is expected to possess the following properties

- It should be rigidly defined
- It should be based on all the items.
- It should not be unduly affected by extreme items.
- It should lend itself for algebraic manipulation.
- It should be simple to understand and easy to calculate

a. STANDARD DEVIATION

It is defined as the positive square-root of the arithmetic mean of the Square of the deviations of the given observation from their arithmetic mean. The standard deviation is denoted by S.D.

The formula for calculating standard deviation is as follows

$$SD = \sqrt{\frac{x^2 - ((\sum x)^2 / n)}{n-1}}$$

Merits

- It is rigidly defined and its value is always definite and based on all the observations and the actual signs of deviations are used.
- As it is based on arithmetic mean, it has all the merits of arithmetic mean.
- It is the most important and widely used measure of dispersion.
- It is possible for further algebraic treatment.
- It is less affected by the fluctuations of sampling and hence stable.
- It is the basis for measuring the coefficient of correlation and sampling.

Demerits

- It is not easy to understand and it is difficult to calculate.
- It gives more weight to extreme values because the values are squared up.
- As it is an absolute measure of variability, it cannot be used for the purpose of comparison.

Example

Standard Deviation (SD)

$$SD = \sqrt{\frac{x^2 - \frac{(\sum x)^2}{n}}{n-1}}$$

$$= \sqrt{\frac{13071.52 - \frac{(878.81)^2}{60}}{60-1}}$$

$$= \sqrt{\frac{13071.52 - 12842.51}{59}}$$

$$= \sqrt{\frac{229.01}{59}} = \sqrt{3.88}$$

$$= 1.97$$

$$\text{Standard deviation} = 1.97$$

b. VARIANCE

The variance is the measures of the variability and it is defined as the average of squared deviation from the mean. The square of the standard deviation is also called variance

$$\text{Variance} = (\text{SD})^2$$

Example

$$\text{Variance} = (\text{SD})^2$$

$$= (1.97)^2$$

$$= 3.88$$

$$\text{Variance} = 3.88$$

c. COEFFICIENT OF VARIATION

The Standard deviation is an absolute measure of dispersion. It is expressed in terms of units in which the original figures are collected and stated. The relative measure is known as the coefficient of variation. The coefficient of variation is obtained by dividing the standard deviation by the mean and expressed in percentage.

$$\text{Coefficient of variation (C.V.)} = \frac{\sqrt{\text{ErrorMS}}}{\text{Grand Mean}} \times 100$$

If we want to compare the variability of two or more series, we can use C.V. The series or groups of data for which the C.V. is greater indicate that the group is more variable, less stable, less uniform, less consistent or less homogeneous. If the C.V. is less, it indicates that the group is less variable or more stable or more uniform or more consistent or more homogeneous.

Example

Coefficient of variation

$$\text{C.V.} = \frac{\sqrt{\text{ErrorMS}}}{\text{Grand Mean}} \times 100$$

$$= \frac{\sqrt{3.015}}{14.63} \times 100$$

$$= \frac{1.74}{14.63} \times 100$$

$$\text{C.V.} = 11.89$$

Coefficient of variation=11.89

d. CRITICAL DIFFERENCE

Critical Difference is obtained by multiplying the SE(d) by at 5% table value for respective error degrees of freedom.

C.D. = SE (d) × t @ 5%

Example

Critical Difference @ 5%

C.D. = SE (d) × t @ 5% (38)

= 1.97 × 1.96

= 3.86

Critical Difference = 3.86

DESIGN OF EXPERIMENT

Tree breeding and improvement is gaining significant attraction and attention across the world in order to improve the genetic resources for productivity and quality improvement. To achieve this goal, new species, varieties and hybrids are identified, evolved and deployed to test verify the superiority of new species/ variety / hybrid/clone over the existing variety or the species currently cultivated. Hence, the breeder should deploy a systematic design of experiment in order to test the varieties identified for superiority compared to existing variety. This needs systematic planning for conducting a field experiment. The establishment of field experiment for evaluation demands a design of experiment. The design of experiment is a method of planning an experiment with a set of rules prior to physical conduct of the experiment to draw a valid conclusion from the experiment.

COMPONENTS OF AN EXPERIMENTAL DESIGN

The following are the components are an experimental design.

i. **Experiment**

It is a systematic procedure of making observations. Otherwise experiment is a means of getting solution to the problem under consideration.

ii. **Treatment**

All object of comparison in the experiment are called as treatments.

Ex. Number of varieties in a breeding trial

Different types of spacing in a silvicultural trial

Different types of chemicals used in pest and disease control experiment

Different types and doses of fertilizers used in nutritional trial

iii. **Experimental unit**

It is the object to which treatments are applied to make observations.

Ex. Plot of a land, laboratories, wood workshop

Pots in pot culture experiment

iv. **Experimental material**

These are the materials in which the whole experiment is planned and performed.

Ex. All biological material deployed in evaluation

v. **Experimental error**

In any experiments the errors are of two types which may be systematic or random. The systematic errors can be explained very easily and the random errors cannot be explained. The errors due to these unexplained variations are said to be experimental errors which may be due to external factor, variation in experimental units, variation due to observations and measurements.

PRINCIPLES OF EXPERIMENTAL DESIGN

The study of experimental design is attributable due to **Prof. Fisher** who defined three basic principles of experimental design which are furnished below

i. Replication

Replication is nothing but the repetition of treatment and comparison. This replication helps to reduce the experimental error and to get a more precise estimate of effect of the various treatments. The replication helps to witness the effect of treatments in each and every experimental unit thereby help to identify the potential of particular treatment. Replication also help to identify the variations occurred due to different treatments.

ii. Randomization

It is the process of allotting the treatment to various experimental units in a random manner. The principle of randomization indicated that every experimental units will have equal chance of receiving any treatment. In the absence of prior knowledge of the variability of the experimental material, the randomization plays a critical role to resolve this issue. It also helps to eliminate the bias and guarantees the elimination of systematic error.

iii. Local control

It is the process of minimizing the experimental error by dividing the experimental area into more homogenous blocks in terms of fertility gradient.

Most commonly Practiced Experimental designs

In most tree breeding and improvement programme, the following are the major experimental designs followed

1. Completely Randomized Design (CRD)
2. Randomized Block Design (RBD)

1. PRACTICAL METHOD OF CRD CALCULATIONS

The replicated wood samples of 18 different tree species were collected and wood density analysis is carried out as per the standard method (TAPPI, 2001). The following data provides the basic density of different tree species in laboratory experiment conducted at FC&RI, Mettupalayam. The practical method of data analysis is furnished below.

S.No	Species	Basic Density (kg / m^3)		
		R1	R2	R3
1.	*Acacia mangium*	567	480	605
2.	*Acrocarpus fraxinifolius*	370	343	340
3.	*Ailanthus excelsa*	416	280	370
4.	*Albizia lebbeck*	345	318	337
5.	*Azardirachta indica*	731	659	553
6.	*Casuarina* sp.	733	705	625
7.	*Dalbergia sissoo*	661	625	536
8.	*Eucalyptus*	659	531	578
9.	*Grevillea robusta*	654	696	789
10.	*Grewia tiliifolia*	524	454	440
11.	*Khaya senegalensis*	559	672	804
12.	*Leucaena leucocephala*	389	382	343
13.	*Melia dubia*	372	435	251
14.	*Neolamarckia cadamba*	275	331	564
15.	*Swietenia macrophylla*	360	423	433
16.	*Tectona grandis*	733	560	666
17.	*Thespesia populnea*	361	350	342
18.	*Toona ciliata*	390	423	420

Calculation

S No	R1	R2	R3	Total	R1^2	R2^2	R3^2	(Total)^2
1	567	480	605	1652	321489	230400	366025	2729104
2	370	343	340	1053	136900	117649	115600	1108809
3	416	280	370	1066	173056	78400	136900	1136356
4	345	318	337	1000	119025	101124	113569	1000000
5	731	659	553	1943	534361	434281	305809	3775249
6	733	705	625	2063	537289	497025	390625	4255969
7	661	625	536	1822	436921	390625	287296	3319684
8	659	531	578	1768	434281	281961	334084	3125824
9	654	696	789	2139	427716	484416	622521	4575321

S No	R1	R2	R3	Total	R1^2	R2^2	R3^2	(Total)^2
10	524	454	440	1418	274576	206116	193600	2010724
11	559	672	804	2035	312481	451584	646416	4141225
12	389	382	343	1114	151321	145924	117649	1240996
13	372	435	251	1058	138384	189225	63001	1119364
14	275	331	564	1170	75625	109561	318096	1368900
15	360	423	433	1216	129600	178929	187489	1478656
16	733	560	666	1959	537289	313600	443556	3837681
17	361	350	342	1053	130321	122500	116964	1108809
18	390	423	420	1233	152100	178929	176400	1520289
Total	**9099**	**8667**	**8996**	**26762**	**5022735**	**4512249**	**4935600**	**42852960**
Mean	**505.5**	**481.5**	**499.8**	**1486.8**	**279040.8**	**250680.5**	**274200**	**2380720**

$$\text{Correction Factor (CF)} = \frac{(\text{Grand Total})^2}{\text{Total No. of Observation}}$$

$$= \frac{(9099+8667+8996)^2}{54}$$

$$= \frac{(26762)^2}{54}$$

$$\text{(CF)} \quad = 13263048.96$$

Treatment Sum of Square

$$\text{Trss} = \frac{\sqrt{Ti^2}}{\text{r}} - \text{CF} \; = \; \frac{42852960}{3} - 13263048.96$$

$$\text{Trss} = 1021271.04$$

Replication Sum of Square

$$\text{RSS} \quad = \frac{(\text{r}1)^2+(\text{r}2)^2+(\text{r}3)^2}{t} - \text{CF } or$$

$$= \frac{\Sigma Ri^2}{t} - \text{CF}$$

$$= \frac{(9099)^2+(8667)^2+(8996)^2}{18} - 13263048.96$$

$$= \frac{82791801+75116889+80928016}{18} - 13263048.96$$

$$= 13268706 - 13263048.96$$

$$\text{RSS} \quad = 5657.04$$

Total Sum of Square

TSS $= R1^2 + R2^2 + R3^2 - CF$ *or*

$= \sum Xi^2 - CF$

= 5022735+4512249+4935600-13263048.96

= 14470584-13263048.96

= 1207535

Error Sum of Square

ESS = Total SS - (TRSS + RSS)

= 1207535 – (1021271.04 + 5657.04) = 80606.92

Analysis of Variance (ANOVA) Table

Source	Degrees of freedom	Sum of square	Mean sum of square	Fcal	Ftab	
					5%	1%
Treatment	t-1 18-1 = 17	1021271.04	60074.77	$\frac{\text{Trss}}{\text{EMSS}}$ $= \frac{60074.71}{5311.97}$ = 11.31	1.933	2.544
Replication	r-1 3-1 = 2	5657.04	2828.52			
Error	(t-1) (r-1) = 17×2 = 34	180606.92	5311.97			
Total		**1207535**				

Calculation for SE d & CD

$$\text{S.Ed} = \frac{\sqrt{2Se^2}}{r} = \frac{\sqrt{2 \times 5311.97}}{3}$$

$$= \sqrt{3541.31}$$

S.Ed = 59.51

CD – 5% level

CD = S.E (d) × + at 5% (34)

t @ 5% = 2.042

CD = 59.51 × 2.042

= 121.51

CD 1% level

CD = S.E (d) × t @ 5% (34)

= 59.51 × 2.457

=146.21

Final table for presentation

From the analyzed data the following table is presented for comparison

S.No	Species	Basic Density (kg / m³)
1.	*Acacia mangium*	550.67
2.	*Acrocarpus fraxinifolius*	351.00
3.	*Ailanthus excelsa*	355.33
4.	*Albizia lebbeck*	333.33
5.	*Azardirachta indica*	647.67**
6.	*Casuarina* Spp.,	687.67**
7.	*Dalbergia sissoo*	607.33
8.	*Eucalyptus*	589.33
9.	*Grevillea robusta*	713.00**
10.	*Grewia tiliifolia*	472.67
11.	*Khaya senegalensis*	678.33**
12.	*Leucaena leucocephala*	371.33
13.	*Melia dubia*	352.67
14.	*Neolamarckia cadamba*	390.00
15.	*Swietenia macrophylla*	405.33
16.	*Tectona grandis*	653.00**
17.	*Thespesia populnea*	351.00
18.	*Toona ciliata*	411.00
	Mean	495.59
	SE.d	59.51
	CD (0.05)	121.51
	CD (0.01)	146.21

Conclusion

Eighteen different tree species were deployed for wood physical quality characterization taking into consideration the basic density as the most prioritized physical wood quality factor. Compared to the average density of all species (**495.59 kg. m^3**), five species viz., *Grevillea robusta* (713.00 kg/m^3), *Casuarina* spp (687.67 kg/m^3)., *Khaya senegalensis* (678.33kg/m^3), *Tectona grandis* (653.00 kg/m^3) and *Azardirachta indica* (647.67 kg/m^3) exhibited significantly higher basic density at 1 % level of significance and these four species are recommended for high density plywood production coupled with other timber product utility.

2. PRACTICAL METHOD OF RBD CALCULATIONS

The following data provides the height of different clones of Kadam obtained through a systematic Clonal Test (CT) established at Forest College and Research Institute, Mettupalayam, Tamil Nadu. The practical method of data analysis is furnished below.

Tree height of various clones of ***Neolamarckia cadamba***

Clones	**Tree height (m)**		
	R1	**R2**	**R3**
AC1	7.45	5.75	5.46
AC 2	8.57	5.29	5.16
AC 3	7.72	6.19	6.37
AC 4	7.88	5.51	5.78
AC 5	7.81	5.61	5.30
AC 6	7.53	5.64	6.15
AC 7	8.76	7.06	6.77
AC 8	7.95	4.67	4.54
AC 9	7.48	5.11	5.38
AC 10	7.43	5.23	4.92
AC 11	7.28	5.39	5.90
AC 12	8.15	5.78	6.05
AC 13	8.35	6.65	6.36
AC 14	8.38	5.10	4.97
AC 15	7.61	5.41	5.10

AC 16	8.33	6.44	6.95
AC 17	7.65	5.95	5.66
AC 18	8.83	5.55	5.42
AC 19	8.00	5.80	5.49
AC 20	5.85	3.96	4.47

Calculation

Clones	R1	R2	R3	Total	R1^2	R2^2	R3^2	Total^2
AC1	7.45	5.75	5.46	18.66	55.50	33.06	29.81	348.20
AC 2	8.57	5.29	5.16	19.02	73.44	27.98	26.63	361.76
AC 3	7.72	6.19	6.37	20.28	59.60	38.32	40.58	411.28
AC 4	7.88	5.51	5.78	19.17	62.09	30.36	33.41	367.49
AC 5	7.81	5.61	5.30	18.72	61.00	31.47	28.09	350.44
AC 6	7.53	5.64	6.15	19.32	56.70	31.81	37.82	373.26
AC 7	8.76	7.06	6.77	22.59	76.74	49.84	45.83	510.31
AC 8	7.95	4.67	4.54	17.16	63.20	21.81	20.61	294.47
AC 9	7.48	5.11	5.38	17.97	55.95	26.11	28.94	322.92
AC 10	7.43	5.23	4.92	17.58	55.20	27.35	24.21	309.06
AC 11	7.28	5.39	5.90	18.57	53.00	29.05	34.81	344.84
AC 12	8.15	5.78	6.05	19.98	66.42	33.41	36.60	399.20
AC 13	8.35	6.65	6.36	21.36	69.72	44.22	40.45	456.25
AC 14	8.38	5.10	4.97	18.45	70.22	26.01	24.70	340.40
AC 15	7.61	5.41	5.10	18.12	57.91	29.27	26.01	328.33
AC 16	8.33	6.44	6.95	21.72	69.39	41.47	48.30	471.76
AC 17	7.65	5.95	5.66	19.26	58.52	35.40	32.04	370.95
AC 18	8.83	5.55	5.42	19.80	77.97	30.80	29.28	392.04
AC 19	8.00	5.80	5.49	19.29	64.00	33.64	30.14	372.10
AC 20	5.85	3.96	4.47	14.28	34.22	15.68	19.98	203.92
Total	**157.01**	**112.09**	**112.20**	**381.30**	**1240.82**	**637.08**	**638.34**	**7328.98**

Correction factor (CF)

$$CF = \frac{(\text{Grand Total})^2}{\text{Total No. of obeservation}}$$

$$CF = \frac{(157.07+112.09+112.20)}{60}$$

$$= \frac{(381.36)^2}{60}$$

$$CF = 2423.92$$

Replication Sum of Square (RSS)

$$RSS = \frac{(r1)^2+(r2)^2+(r3)^2}{t} - CF$$

$$= \frac{((157.07)^2+(112.09)^2+(112.20)^2)}{20} - CF$$

$$= \frac{24670.98+12564.17+12588.84}{20} - 2423.92$$

$$= 2491.20-2423.92$$

$$RSS = 67.28$$

Treatment Sum of Square (Trss)

$$Trss = \frac{\Sigma Ti^2}{r} - CF$$

$$= \frac{7328.98}{3} - 2423.92$$

$$Trss = 19.07$$

Total Sum of Square (TSS)

$$TSS = (1240.82+637.08+638.34) - 2423.92$$

$$= 2516.24 - 2423.92$$

$$TSS = 92.32$$

Error Sum of Square (ESS)

$$ESS = TSS - (Trss+RSS)$$

$$= 92.32 - (19.07+67.28)$$

$$= 92.32 - 86.35$$

$$ESS = 5.97$$

Analysis of variance (ANOVA) Table

Source	Degrees of freedom	Sum of square	Mean sum of square	Fcal	Ftab	
					5%	1%
Replication	2	67.28	33.64	33.64	1.84	2.07
Treatment	19	19.07	1.00	6.37		
Error	38	5.97	0.1571			
Total		**92.32**				

Calculation of S.Ed & CD

$$\text{S.Ed} = \frac{\sqrt{2Se^2}}{r} - \text{CF} = \frac{\sqrt{2\times0.1571}}{3}$$

$$= \frac{\sqrt{0.3142}}{3} = 0.3236$$

CD 5% level

CD = SE (d) × t @ 5% (38)

= 0.3236 × 2.021 = 0.6540

CD 1% level

CD = SE (d) × t @ 1% (38)

= 0.3236 × 2.423 = 0.7840

Final table for presentation

From the analyzed data, the following table is presented for comparison

Clones	Tree height (m)
AC1	6.22
AC 2	6.34
AC 3	6.76
AC 4	6.39
AC 5	6.24
AC 6	6.44
AC 7	7.53**
AC 8	5.72
AC 9	5.99

Clones	Tree height (m)
AC 10	5.86
AC 11	6.19
AC 12	6.66
AC 13	7.12*
AC 14	6.15
AC 15	6.04
AC 16	7.24**
AC 17	6.42
AC 18	6.6
AC 19	6.43
AC 20	4.76
Mean	**6.36**
SE.d	**0.32**
CD (0.05)	**0.65**
CD (0.01)	**0.78**

Conclusion

Since there is no released variety available in Kadam, the tested clones are compared with the average height of all clones. Compared to the average height (6.36 m), only 3 clones viz., AC 7 (7.53 m), AC 13 (7.12 m) and AC 16 (7.24 m) recorded significantly higher tree height and hence these three clones are proved to be superior and may be advanced for amplified clonal test.

SUMMARY

Application of statistics is a prerequisite in tree breeding and improvement programme with deals with collection of data, processing of data, analysis, presentation and interpretation of processed data. Several elementary statistic methods like Arithmetic Mean (AM), Median, Mode, Standard Deviation (SD), Variance, Coefficient of Variation, Critical Difference and test of significance are very important for estimation and interpretation of growth and biometric attributes derived from tree improvement programme. Similarly, the designs of experiment are also very important in order to identify the treatmental effect on the experimental materials. This chapter summarized all these basic and applied statistical tools with relevant calculations and estimations and are presented with suitable conclusions. This chapter may be of great use to the young researchers, students, tree breeders and professional foresters for direct use in ongoing tree breeding and improvement programme of wide range of tree species.

8

BIOMETRIC TECHNIQUES IN TREE IMPROVEMENT

INTRODUCTION

The tree breeding and improvement aimed to enhance the yield components depend primarily on quantification of yield and quality traits. The yield and quality parameters exhibited wider variation which may be due to genetic or due to environment or due to combined effect of genotype and environmental interactions. Under such circumstances, the primary role of the tree breeder is to assess and determine the estimates of genetic and environmental components. This estimation will help the breeder to manipulate the breeding and environmental components in order to achieve the objective envisage. Hence the estimation of various genetic estimates like genetic and phenotypic coefficient of variation, heritability and the associated genetic gains are important genetic tools which will help the breeder to determine the efficiency of a selection and breeding program. To understand the causes of variations apportioning the total phenotypic variation into phenotypic coefficient of variation (PCV) and genotypic coefficient of variation (GCV) is very important which will indicate the variability estimates due to genetics and/or due to environment. Similar to variability, heritability estimate has also play an important role in tree breeding because it acts as an index of relative strength of inheritance and environment. Similarly, the genetic gain indicates the predominance of additive gene action and their estimate also extent greater support to tree breeders. All these aspects indicated above, clearly states that the tree breeding and improvement program need to be tested for various genetic tools in order to identify the success or failure of the tree breeding program. Keeping this in view, the current chapter on biometric techniques in tree improvement has been designed and presented with actual field data for easy understanding.

DEFINITION OF TERMINOLOGIES

Variation	:	Variation is the genetic differences among trees and the interactions between the tree genotypes and the environments in which they grow.
Heritability	:	Heritability is the ratio between the genotypic variance to the phenotypic variance.
Narrow sense heritability	:	Narrow sense heritability is the ratio of additive Genetic variance to the total phenotypic variance Narrow sense heritability is the proportion of total variance that is attributable to the additive effects of genes.
Broad sense heritability	:	Broad sense heritability is the ratio of total genetic variance (additive + non additive) to the total phenotypic variance.
Selection differential	:	The difference between the mean phenotypic value of the selected plants and the mean phenotypic value of the parental population is termed as selection differential.
Genetic gain	:	The difference between the mean phenotypic value of the progeny of selected plants and the base or parental population is known as genetic gain.
Genetic advance	:	Improvement in the mean genotypic value of selected plants over the parental population is known as genetic advance.

VARIATION

- Variation is the genetic differences among trees and the interactions between the tree genotypes and the environment in which they grow. Some variations are predictable and others are non predictable by the breeder.
- Variation is caused by genetic makeup and is called as genetic variation and the variation is due to environment it is called environmental variation.

CATEGORIES OF VARIATION

The variation present in trees are categorized into five types and they are:

i) **Geographic or provenance variation:** Geographic or provenance differences are significantly large especially the characters which are highly adaptable. This variation is used by the breeders for any tree improvement programme.

ii) **Variability among sites within provenance:** The given provenance can exhibit significant differences among themselves due to the sites where they have been introduced. These variations are not genetically fixed but represent the effect of varied site and environmental conditions on the tree growth and development.

iii) **Variation between stands within sites:** Very often the stands or plantations within a site exhibit wider variation which may be due to the localized site or natural variation found within a tree. This is very important particularly for tree form and structure.

iv) **Variation between individual trees within stand:** Sometime, within a stand the individual trees exhibit great variation which may be genetic variation or in some times it may be due to the local soil type. Many individual tree characteristics are strongly under tree control and the breeders can exploit this variation through selection and incorporate them in their breeding programme.

v) **Variation within trees**: Trees exhibit variation within themselves for certain wood quality traits. For eg. Wood density, Wood specific gravity, Wood Cellulose etc. will differ based on the tree position i.e. base, middle and top of the tree. Within tree variation is not genetically controlled but this variation is of importance for biometric assessments and wood sample collection.

IMPORTANCE OF VARIATION

- The variation exists in a population will help the breeder to produce more quickly the desired products with a greater uniformity.

- Variation will be useful to identify the best genotypes suitable for varied utility viz., timber, pulpwood, plywood, match, veneer, sports goods etc.
- Variation will be useful to identify the highly adaptable variety for extreme weather conditions like cold or drought or flood.
- Variation is essential to screen and identify trees resistance towards pest and diseases.
- Variation is useful to identify high yielding varieties.

TYPES AND QUANTIFICATION OF VARIATION

Variation that exist in tree population is primarily due to genetic or environmental or combination of genetic and environment. The combined action of genotype and environment decides the phenotypes of the population and is described as follows:

Phenotype = Genotype + Environment

The phenotypic variation is due to the combined action of genetic and environmental variation which is described a s follows

Phenotypic variation = Genetic variation + Environmental variation

Symbolically it is expressed as $\sigma P^2 = \sigma G^2 + \sigma E^2$

Determination of Genetic Estimates

In any tree improvement programme, the determination of following genetic estimate are very important in order to assess the causes and extent of variation, characters with strong heritability and the resultant genetic gain. Accordingly, the following genetic tests are demonstrated with suitable example for the benefit of the readers of the book.

I. Variability estimation

II. Heritability Estimation

III. Assessment of Genetic gain

I. ESTIMATION OF VARIATION

The quantification of variation especially genotypic coefficient of variations and phenotypic co-efficient of variations are expressed through a field data obtained from a *Melia dubia* seed sources evaluation tests.

The following data provides the height of different progenies of *Melia dubia* obtained through a systematic Seed Source Evaluation Test established at Forest College and Research Institute, Mettupalayam, Tamil Nadu. The practical method of data analysis is furnished below.

The height of 20 seed sources of *Melia dubia* established though a standard seed source evaluation trail (with three replication) was measured using altimeter and expressed in centimeter. This height as a trait has been used for estimation of variability (Genotypic co-efficient of variation & Phenotypic co-efficient of variation), Heritability and Genetic advance. The procedure for determining various genetic estimates are furnished below.

Tree height Assessment of *Melia dubia* seed sources test

S.No	Seed Sources	R1	R2	R3
1.	MD 1	14.72	17.33	13.50
2.	MD 2	14.94	17.46	16.48
3.	MD 3	15.33	13.78	15.39
4.	MD 4	9.94	13.83	17.33
5.	MD 5	11.73	18.11	15.50
6.	MD 6	13.27	16.44	13.11
7.	MD 7	11.17	14.50	14.22
8.	MD 8	12.52	17.33	16.28
9.	MD 9	14.09	12.06	15.67
10.	MD 10	11.13	15.78	13.17
11.	MD 11	13.76	16.18	16.53
12.	MD 12	13.80	12.64	14.30
13.	MD 13	14.73	16.18	15.89
14.	MD 14	15.39	18.50	14.22
15.	MD 15	14.53	16.56	15.50
16.	MD 16	14.71	16.61	12.36
17.	MD 17	14.06	14.50	9.52
18.	MD 18	12.83	16.94	16.07
19.	MD 19	11.78	15.44	16.78
20.	MD 20	13.17	14.56	13.33

Calculations and Estimation

S.No	Seed Source	R1	R2	R3	Total	$(R1)^2$	$(R2)^2$	$(R3)^2$	$(Total)^2$
1.	MD 1	14.72	17.33	13.50	45.56	216.74	300.44	182.25	2075.31
2.	MD 2	14.94	17.46	16.48	48.88	223.34	304.89	271.52	2389.58
3.	MD 3	15.33	13.78	15.39	44.50	235.11	189.83	236.82	1980.25
4.	MD 4	9.94	13.83	17.33	41.11	98.89	191.36	300.44	1690.12
5.	MD 5	11.73	18.11	15.50	45.34	137.54	328.01	240.25	2055.61
6.	MD 6	13.27	16.44	13.11	42.82	176.00	270.42	171.90	1833.74
7.	MD 7	11.17	14.50	14.22	39.89	124.69	210.25	202.27	1591.12
8.	MD 8	12.52	17.33	16.28	46.13	156.81	300.44	264.97	2128.28
9.	MD 9	14.09	12.06	15.67	41.81	198.50	145.34	245.44	1748.17
10.	MD 10	11.13	15.78	13.17	40.09	123.95	249.46	173.36	1607.56
11.	MD 11	13.76	16.18	16.53	46.47	189.22	261.90	273.17	2159.15
12.	MD 12	13.80	12.64	14.30	41.04	190.44	167.56	204.49	1684.65
13.	MD 13	14.73	16.18	15.89	46.81	217.07	261.90	252.46	2190.76
14.	MD 14	15.39	18.50	14.22	48.11	236.82	342.25	202.27	2314.68
15.	MD 15	14.53	16.56	15.50	46.59	211.22	274.09	240.25	2170.52
16.	MD 16	14.71	16.61	12.36	43.68	216.42	275.93	152.66	1907.75
17.	MD 17	14.06	14.50	9.52	38.08	197.56	210.25	90.67	1449.92
18.	MD 18	12.83	16.94	16.07	45.84	164.69	287.11	258.14	2101.71
19.	MD 19	11.78	15.44	16.78	44.00	138.72	238.53	281.49	1936.00
20.	MD 20	13.17	14.56	13.33	41.06	173.36	211.86	177.78	1685.56
	TOTAL	**267.61**	**315.07**	**295.14**	**877.81**	**3627.09**	**5021.83**	**4422.60**	**38700.46**

$$\text{Correction Factor (CF)} = \frac{(\text{Grannd Total})^2}{\text{Total No. of Observation}}$$

$$\text{Correction Factor (CF)} = \frac{(267.61+315.07+295.14)^2}{60}$$

$$= \frac{(877.81)^2}{60}$$

$$= \frac{770552.35}{60}$$

$$\text{CF} = 12842.54$$

Treatment Sum of Square (Trss)

$$\text{Trss} = \frac{\Sigma Ti^2}{r} - \text{CF} = \frac{38700.46}{3} - 12842.54$$

$$\text{Trss} = 12900.15 - 12842.54$$

$$\text{Trss} = 57.61$$

Replication sum of Square (RSS)

$$\text{RSS} = \frac{(r1)^2+(r2)^2+(r3)^2}{t} - \text{CF} \text{ or } \frac{Eri^2}{t} - \text{CF}$$

$$= \frac{(267.61)^2+(315.07)^2+(295.14)^2}{20} - 12842.54$$

$$= \frac{71612.73+99267.00+87106.96}{20} - 12842.54$$

$$= \frac{257986.70}{20} - 12842.54$$

$$= 12899.34 - 12842.54$$

$$\text{RSS} = 56.84$$

Total Sum of Square (TSS)

$$\text{TSS} = (R1^2+R2^2+R3^2) - \text{CF}$$

$$= (3627.09+5021.83+4422.60) - 12842.54$$

$$= 13071.52 - 12842.54$$

$$\text{TSS} = 288.98$$

Error Sum of Square (ESS)

ESS = Total SS – (Trss + RSS)

= 228.98 – (57.61 + 56.80)

= 228-98 – 114.41

ESS = 114.57

Analysis of Variance (ANOVA) Table

Source	Degrees of freedom	Sum of Square	Mean sum of Square
Treatment	t-1 20-1 = 19	57.61	3.032 (M1)
Replication	r-1 3-1 = 2	56.80	28.4
Error	(t-1) (r-1) = 19 × 2=38	114.57	3.015 (M2)
Total		**228.98**	

Genetic Variance (GV)

$$\text{Genetic Variance (GV)} = \frac{\text{M1–M2}}{\text{r}}$$

$$= \frac{3.032\text{-}3.015}{3}$$

$$= \frac{0.017}{3}$$

$$\text{GV} = 0.00567$$

$$\text{Mean} = \frac{E \times i}{r} = \frac{877.81}{60}$$

$$\text{Mean} = 14.6301$$

Genotypic Co-efficient of Variation (GCV)

$$\text{GCV} = \frac{\sqrt{\text{GV}}}{\text{mean}} \times 100$$

$$= \frac{\sqrt{0.00567}}{14.6301} \times 100$$

$$\textbf{GCV} = \frac{0.075}{14.6301} \times 100$$

$$= 0.5126$$

Environment Variance (Error Variance) = M2 = 3.015

EV = 3.015

Phenotypic Variance (PV)

PV = GV + EV

PV = 0.00567 + 3.015 = 3.02067

Phenotypic Co-efficient of Variation (PCV)

$$\text{PCV} = \frac{\sqrt{\text{PV}}}{\text{mean}} \times 100$$

$$= \frac{\sqrt{3.02067}}{14.63.01} \times 100 = \frac{1.73810}{14.6301} \times 100$$

PCV = 11.87

Conclusion

In the above variability estimation, It is very clear that the PCV value is higher (11.87) than its corresponding genotypic value (0.5126) which indicate that the characters are greatly influenced by the environment.

II. HERITABILITY

It is an important genetic tool which indicates the degree to which the parents transfer their characteristics to the off springs.

Types of Heritability

There are two types of heritability depending on the type of genetic variance utilized for tree improvement.

a) **Narrow sense heritability:** It is the ratio of additive genetic variance to the total phenotypic variance or it is the proportion of total variance that is attributable to the additive effects of genes. It is described as

$$h^2 = \sigma_A^2 / (\sigma_A^2 + \sigma_{NA}^2 + \sigma_E^2)$$

b) **Broad sense heritability:** It is the ratio of total genetic variance (additive + non additive) to the total phenotypic variance.

$$H^2 = (\sigma_A^2 + \sigma_A^2) / (\sigma_A^2 + \sigma_{NA}^2 + \sigma_E^2)$$

Applications of heritability

- Heritability helps the breeder to select individuals with high heritability.
- It helps to predict the genetic gain realized through tree improvement.

Quantification of Heritability

$$\text{Heritability (\%)} = \left(\frac{GV}{PV}\right) \times 100$$

$$= \left(\frac{0.00567}{3.02067}\right) \times 100$$

Heritability = 0.1877

Conclusion

The heritability value estimated in the above calculation is of low value which indicates that the height as a character is not reliable for selection in Melia.

III. GENETIC GAIN

Genetic gain is the product of heritability and selection differential which is described as

Genetic gain = $h^2 \times Sd$

Where, h^2 is the heritability

Sd is the selection differential

Application

- Genetic gain helps to assess the tree improvement process towards improving the quality and quantity of a product.
- Higher genetic gain indicates the positive aspect of tree improvement programme
- Lower genetic gain indicates the failure in breeding process.

GENETIC ADVANCE

Genetic advance is an important parameter which helps the breeder in selection programme along with heritability. It refers to the improvement made in mean genotypic value of selected individuals over the parental population. The existence of variability, heritability and selection intensity may influence the genetic advance.

Significance of Genetic advance

- The genetic advance indicates that if the value is higher which may be due to the characters are governed by additive gene action and selection for such traits will lead to very good improvement.
- Suppose if the genetic advance value is low, it means that the character is governed by non-additive gene action and the selection for that trait may be avoided.

Quantification of Genetic Advance

$$\text{Genetic Advance} = h^2 \times \propto P \times K$$

$$SD(\propto P) = \frac{\sqrt{EX^2} - (EX)^2/n}{n-1}$$

$$= \frac{\sqrt{13071.52} - \frac{(877.81)^2}{90}}{60-1}$$

$$= \frac{\sqrt{13071.52} - 8561.67}{59} = \frac{\sqrt{4509.85}}{59}$$

$$= \sqrt{76.44}$$

$$\propto P = 8.7428 \textit{ or } SD = 8.7428$$

$$= h^2 \times \propto P \times K$$

$$= 0.1877 \times 8.7428 \times 2.06$$

$$= 3.3805$$

Genetic Advance = 3.3805

$$GA\% = \frac{GA}{\text{mean}} \times 100$$

$$= \frac{2.749}{14.6301} \times 100$$

$$= 0.1879 \times 100$$

$$\textbf{GA\%} = \textbf{18.79\%}$$

Final Table

S. No	Genetic Estimates	Values
1.	GCV	0.5126
2.	PCV	11.87
3.	Heritability	0.1877
4.	Genetic Advance as Mean	18.79

Conclusion

The genetic advance estimation in the above study exhibited moderate value and the heritability is also very low. Hence low genetic advance coupled with low heritability indicates that the character is highly influenced by environmental effect and the selection process deploying height as a trait may not be effective for improvement programme.

SUMMARY

The growing interest on plantation forestry and Agroforestry have attracted genetic improvement of wide range of tree species for higher productivity and improved wood quality. It is very well witnessed that genetically improved plantations could produce higher utilizable biomass and reduces the wastages significantly during the wood processing. Hence tree improvement is vital for the ongoing plantation establishment programme. This tree breeding strategy is largely depends upon the extent of variability in a base population which is usually measured by various parameters like genotypic and phenotypic variance and the associated genotypic and phenotypic coefficient of variances. In the tree breeding programme, it is very important that genetic component of the variation is the only significant factor to be transmitted to next generation. Hence the estimation of heritability has gained significance. The heritability guides the tree breeder in selection programme and it is very well established that the best gains are made for characteristics that are under strong genetic control. Similarly the genetic gain is also a very important parameter and this along with the heritability guide the breeder in efficient tree improvement programme. Hence the current chapter has been conceived and presented with suitable field level examples. All the genetic estimates like PCV, GCV, heritability and genetic advance are demonstrated with actual data and presented through sustainable conclusions. This will be more useful to graduating scholars and post graduating researchers towards understanding and applying the genetic tests towards improving wide range of species.

9

GENETIC ASSOCIATION

INTRODUCTION

Tree improvement of a given species leads to improvement in yield as well as quality parameters. The yield is a component of genetic improvement and the associated silvicultural management practices. This yield depends on multiple traits like height, diameter, stem form, clear bole height etc., which either directly or through other characters contributes towards yield improvement. Hence it is a complex phenomenon which depends on the relationship between several variables which could be assessed through the correlation analysis. Correlation is estimated through genotypic and phenotypic correlation wherein the former one is more reliable which is controlled by genetic association and the later one is influenced by environment. The genotypic correlation is an estimated values where as the phenotypic correlation is an estimated value from the genotype and environmental interaction. Hence genotypic correlation is a more reliable estimate towards assessing the degree of relationship between different characters. The association study in most tree improvement programme is gaining more significance in the recent past due to its role in examining the relationship between different characters towards achieving yield improvement or wood quality improvement. Hence the current chapter on association studies has been designed and presented with suitable practical demonstration deploying actual field data to benefit the readers of the book.

CORRELATION - DEFINITIONS

- Correlation is the study of relationship between two or more quantitative variables. (Dhamu and Ramamoorthy, 2007)
- The measures of the degree of relationship between two continuous variables are called correlation coefficient.
- Correlation is a statistical measure which is used to find out the degree and direction of relationship between two or more variables.
- Correlation coefficient analysis measures the mutual relationship between various characters and determines the components characters in which selection can be based for genetic improvement in productivity. (Phundan Singh and Narayanan,1993)

PROPERTIES OF CORRELATION CO-EFFICIENT

i. The correlation coefficient value ranges between –1 and +1.

ii. It is independent of the unit of measurement

iii. It is measures the degree and direction of association between two or more variables.

iv. The degree of relationship between two variables is symmetric

v. The two variables are independent, then the correlation coefficient is zero.

vi. The correlation coefficient is independent of the change of origin and scale.

vii. The coefficient of correlation is the geometric mean of two regression coefficient.

TYPES OF CORRELATION

There are three major types of correlation.

1. Simple correlation
2. Partial Correlation
3. Multiple correlation

1. Simple correlation

The relationship between two characters is a simple correlation. It is also known as total correlation or zero correlation.

The simple correlation is divided into three types.

i. Phenotypic correlation

ii. Genotypic correlation

iii. Environment correlation

i. Phenotypic correlation

- The relationship between two variables directly observed is phenotypic correlation.
- Phenotypic correlations include both genotypic and environmental effects and differ under various environmental conditions.

ii. Genotypic correlation

- The inherent or heritable association between two characters is genotypic correlation.
- It is also known as genetic correlation
- The relationship between two traits remains the same (+ or –) in the parental population and segregation population, it means the association is mainly due to pleiotropy
- This type of correlation is more stable

iii. Environment correlation

- Environmental correlation is mainly due to the environment effect.
- It is also knows as error variance
- It is not important for any tree breeder because it is not heritable and stable.

2. Partial correlation

The correlation between two variables is measured by eliminating the effect of third variable is partial correlation. It is also known as net correlation. The partial correlation is a study of relationship between one dependent and one independent variable and other one independent variable is ignored. The partial correlation is divided into two types

i. First order partial correlation

ii. Second order partial correlation

i. First order partial correlation

- First order partial correlation involves three variables at a time.
- It measures the association between two variables by eliminating the effect of third variable.
- It is denoted by $r_{12.3}$
- Its significance is tested against n-3 degrees of freedom

ii. Second order partial correlation

- At a time, Second order partial correlation measures the relationship between four variables.
- It measures the association between two variables by eliminating the effect of third and forth variable.
- It is denoted by $r_{12.34}$
- Its significance is tested against n-4 degrees of freedom

3. Multiple correlation

- The estimate of joint influence of two or more independent variables on a depend variable is called multiple correlation.
- It involves three or more variables at a time. In partial correlation measures, the relationship of two variables by making third variable as constant but multiple correlation the effect of all the independent variables is analysed on a dependent variable.
- The multiple correlation value is always higher than simple and multiple correlations
- It is a non negative estimates and it can never be negative.
- The correlation values ranged between 0 and 1

PRACTICAL METHOD OF CORRELATION ESTIMATION

The following data provides the biometric attributes of different progenies of *Melia dubia* obtained through a systematic Seed Source Test (SST) established at Forest College and Research Institute, Mettupalayam, Tamil Nadu. The practical method of data analysis is furnished below.

Biometric attributes of various seed sources of *Melia dubia*

S. No	Name of the seed source	Height (m)	Girth (cm)	Volume (m^3)
1.	MD 1	15.19	53.77	0.213
2.	MD 2	16.29	55.52	0.239
3.	MD 3	14.83	54.00	0.213
4.	MD 4	13.70	50.44	0.179
5.	MD 5	15.11	53.68	0.221
6.	MD 6	14.27	50.33	0.177
7.	MD 7	13.30	44.54	0.131
8.	MD 8	15.38	51.76	0.199
9.	MD 9	13.94	50.58	0.184
10.	MD 10	13.36	47.10	0.147
11.	MD 11	15.49	53.73	0.216
12.	MD 12	13.68	45.97	0.138
13.	MD 13	15.60	57.33	0.252
14.	MD 14	16.04	54.02	0.237
15.	MD 15	15.53	50.52	0.189
16.	MD 16	14.56	48.61	0.181
17.	MD 17	12.69	47.30	0.153
18.	MD 18	15.28	53.46	0.217
19.	MD 19	14.67	53.35	0.218
20.	MD 20	13.69	47.22	0.146

Calculations and Estimation

S. No	Seed sources	Height	Girth	Volume	Height^2	Girth^2	Volume^2	H XG	H XV	G XV
1	MD 1	15.19	53.77	0.213	230.59	2891.65	0.045	816.57	3.23	11.45
2	MD 2	16.29	55.52	0.239	265.51	3082.31	0.057	904.64	3.90	13.27
3	MD 3	14.83	54.00	0.213	220.03	2915.60	0.045	800.95	3.16	11.50
4	MD 4	13.70	50.44	0.179	187.79	2544.227	0.032	691.22	2.46	9.04
5	MD 5	15.11	53.68	0.221	228.40	2881.70	0.049	811.29	3.34	11.87
6	MD 6	14.27	50.33	0.177	203.75	2532.70	0.031	718.36	2.53	8.93
7	MD 7	13.30	44.54	0.131	176.79	1984.21	0.017	592.28	1.74	5.83
8	MD 8	15.38	51.76	0.199	236.48	2678.64	0.040	795.89	3.06	10.31
9	MD 9	13.94	50.58	0.184	194.24	2558.11	0.034	704.90	2.56	9.30
10	MD 10	13.36	47.10	0.147	178.62	2218.76	0.022	629.53	1.96	6.91
11	MD 11	15.49	53.73	0.216	239.91	2887.27	0.047	832.27	3.35	11.63
12	MD 12	13.68	45.97	0.138	187.18	2113.27	0.019	628.94	1.89	6.34
13	MD 13	15.60	57.33	0.252	243.42	3287.11	0.063	894.51	3.93	14.44
14	MD 14	16.04	54.02	0.237	257.19	2918.40	0.056	866.36	3.80	12.80
15	MD 15	15.53	50.52	0.189	241.17	2552.49	0.036	784.59	2.94	9.56
16	MD 16	14.56	48.61	0.181	211.97	2363.04	0.033	707.74	2.63	8.79
17	MD 17	12.69	47.30	0.153	161.10	2236.94	0.023	600.31	1.94	7.24
18	MD 18	15.28	53.46	0.217	233.52	2857.50	0.047	816.88	3.32	11.60
19	MD 19	14.67	53.35	0.218	215.11	2846.42	0.048	782.49	3.20	11.65
20	MD 20	13.69	47.22	0.146	187.28	2229.59	0.021	646.19	2.00	6.90
	Mean	**14.63**	**51.16**	**0.193**	**215.00**	**2629.00**	**0.038**	**751.30**	**2.85**	**9.97**
	Total	**292.60**	**1023.23**	**3.851**	**4300.05**	**52579.98**	**0.766**	**15025.91**	**56.94**	**199.35**

Correlation's denoted as "r"

$$r = \frac{SPxy}{\sqrt{SSx.SSy}}$$

$$SPxy = \Sigma xy - \frac{\Sigma x \times \Sigma y}{n}$$

$$SSx = \Sigma x^2 - \frac{(\Sigma x)^2}{n}$$

Correlation between Height and Girth

$$\text{SPHG} = \Sigma HG - \frac{\Sigma H \cdot \Sigma GG}{n}$$

$$= 15025.91 - \frac{292.60 \times 1023.23}{20}$$

$$= 15025.91 - 14969.85$$

$$= 56.06$$

$$\text{SPHG} = \Sigma H2 - \frac{(\Sigma H)^2}{n}$$

$$= 4300.05 - \frac{(292.60)^2}{20}$$

$$= 4300.05 - 4280.74$$

$$= 19.31$$

$$\text{SSG} = \Sigma G - \frac{(\Sigma G)^2}{n}$$

$$= 52579.98 - \frac{(1023.23)^2}{20}$$

$$= 52579.98 - 52349.98$$

$$= 230.00$$

$$r = \frac{SPHG}{\sqrt{SSH \cdot SSG}} = \frac{56.06}{\sqrt{19.31 \times 230}} = \frac{56.06}{66.64}$$

$$r = 0.841$$

Significance / non significance

r = 0.841

n = 20

$$t = \frac{r}{\sqrt{1-r^2}} \times \sqrt{n}-2$$

$$= \frac{0.841}{\sqrt{1-0.841^2}} \times \sqrt{20}-2$$

$$= \frac{0.841}{\sqrt{0.29}} \times \sqrt{18}$$

$$= \frac{0.841}{0.538} \times 4.24$$

$= 6.62$

t Calculated value = 6.62

t Tabulated value = 2.10

t c > t t

Since, the calculated value (6.62) is higher than the tabulated value (2.10) the correlation existed between height and girth is highly significant

Correlation between Height & Volume

$$\text{SPHV} = \sum HV - \frac{\sum H \times \sum V}{n}$$

$$= 56.94 - \frac{292.60 \times 3.851}{20}$$

$= 56.94 - 56.34$

$= 0.60$

$$\text{SSH} = \sum H2 - \frac{(\sum H)^2}{n}$$

$$= 4300.05 - \frac{(292.60)^2}{20}$$

$= 19.31$

$$SSV = \sum V2 - \frac{\sum V^2}{n}$$

$$= \sum 0.766 - \frac{(3.851)^2}{20}$$

$$= 0.766 - 0.741$$

$$= 0.025$$

$$r = \frac{SPHV}{\sqrt{SSH \cdot SSV}} = \frac{0.60}{\sqrt{19.31 \times 0.025}}$$

$$= \frac{0.60}{0.6948}$$

$$r = 0.863$$

Significance / non significance

r = 0.869

n = 20

$$t = \frac{r}{\sqrt{1-r^2}}$$

$$= \frac{0.863}{\sqrt{(1-0.863)^2}} \times \sqrt{18}$$

$$= \frac{0.863}{0.504} \times 4.24$$

t c = 7.260

t Calculated value = 7.260

t Calculated value = 2.10

t c > t t

Since, the calculated value (7.260) is higher than the tabulated value (2.10) the correlation existed between height and volume is highly significant.

Correlation between Girth & Volume

$$SP\ GV = \sum GV - \frac{\sum G . \sum V}{n}$$

$$= 199.35 - \frac{1023.23 \times 3.851}{20}$$

$$= 195.35 - 197.02$$

$$= -1.67$$

$$\text{SSG} = \sum G - \frac{(\sum G)^2}{n}$$

$$= 52579.98 - \frac{(1023.23)^2}{20}$$

$$= 230.00$$

$$\text{SSV} = \sum V2 - \frac{(\sum V)^2}{n}$$

$$= 0.760 - \frac{(3.85)^2}{20}$$

$$= 0.025$$

$$\text{r} = \frac{SPGV}{\sqrt{SSH} \cdot SSV} = \frac{-1.67}{\sqrt{230} \times 0.025}$$

$$= \frac{-1.67}{2.40}$$

$$\mathbf{r = 0.696}$$

Significant / non significant

$$\text{t} = \frac{r}{\sqrt{1} - r^2} - \sqrt{n} - 2$$

$$\text{r} = -0.696$$

$$\text{n} = 20$$

$$= \frac{-0.696}{\sqrt{1} - (-0.696)^2} \times \sqrt{18}$$

$$= \frac{-0.696}{0.718} \times 4.24$$

t c = - 4.11

t c Calculated value = -4.11

t t tabulated value = 2.10

don ' t consider +/-

t c > t t

The correlation coefficient does not take into consideration + or –. Hence the calculated value of 4.11 (Though it is -4.11) is greater than the tabulated value (2.10) and the association between girth and volume is significant.

Correlation Table

Parameter	Height	Girth	Volume
Height	1.000	0.841*	0.863*
Girth		1.000	-0.696*
Volume			1.000

Inference

The correlation analysis of biometric attributes collected from Melia progeny test indicated that height exhibited significant and positive correlation with volume (0.863) and attests that height is a contributing factor for volume improvement of *Melia dubia*. Similarly height significantly influenced volume through inter association with girth (0.841). In a holistic analysis, height and girth had influence on volume improvement and hence these two parameters can be considered in Melia improvement programme. However in the current analysis, girth exhibited negative but significant association with volume (-0.696) and also extend the scope of its application in Melia improvement programme.

PATH ANALYSIS

Path analysis is a genetic association tool which apportions the correlation coefficient into direct and indirect effects and measures the relative importance of causal factors involved.

Importance

- It is more useful tool to study the effect of direct and indirect effects on the traits of interest.
- It is extensively useful to understand the complex traits in tree breeding programmes.

PRACTICAL METHOD OF PATH ANALYSIS

The data generated from Melia seed source test used in correlation analysis is also deployed in path analysis. The results of path analysis are furnished below for easy understanding and to draw valid conclusion.

Direct (Diagonal) and indirect effects of growth traits of *Melia dubia* seed sources

Characters	Plant height	Girth at breast height
Plant height	0.2090	0.0052
Girth at breast height	0.1185	0.0092

Residual Effect: 0.4236

Inference

The path analysis indicated that all the growth attributes exerted positive direct effect on volume. The maximum positive direct effect was exerted by plant height (0.2090) followed by girth at breast height (0.0092). The girth at breast height exerted its maximum positive indirect effect via plant height (0.1185) on volume of *Melia dubia*. The height exerted a minimal but positive indirect effect through girth at breast height (0.0052) on volume. The study indicated that both height and GBH are contributing factors towards achieving maximum volume both directly and indirectly.

SUMMARY

Tree breeding and improvement programme in most tropical tree species depends predominantly on the selection process which formed a most important activity of all successful tree breeding activities. The tree yields particularly the tree volume is a complex and highly variable character which is influenced by several component characters like height, girth at breast height, tree form, clear bole height, basal girth etc., In the integrated structure of plants, the overall correlation observed between two variables is a function of series of direct and indirect relationships between different variables. This has been demonstrated using height, GBH and volume as variable and their direct and indirect relationship has been demonstrated with problem solving example. Similarly the path analysis which gives an insight into a complex relationship between different characteristic of a tree based biological system. It also apportions the correlation coefficient into direct and indirect effects and helps in understanding the complex traits of a tree breeding system. The path analysis has been demonstrated with suitable example deploying the actual data collected from a Melia progeny test. The role of growth attributes in extending direct effects on volume and also indirectly through other characters has been demonstrated. These demonstrations of correlation and path analysis will act as a readymade reference to the graduating scholars, tree breeders and scientists in order to apply them in their respective tree breeding programme to know the magnitude and type of association that exists between growth and the associated components. Similarly the readers could understand the role of complex traits in tree breeding programme through path analysis and also understand the interrelationship of characters. This will also help the tree breeders in identification of effective components to be deployed in tree breeding programme in the prioritized species.

10

SEED ORCHARDS

INTRODUCTION

The growing importance on promotion of agroforestry, plantation forestry and industry-based agroforestry activities coupled with environmental plantings have demanded huge volume of quality planting materials. The plantation establishment and the associated success of plantation depend heavily on quality of planting materials used. Today, the planting materials are produced both through seedlings and clones. The current estimate in the country indicates that over 70% of the seedling produced in the country is seed based barring few exceptions like industrial agroforestry plantations established by pulp and paper industries who predominantly deploy clonal plants. Under such circumstances, seed-based seedling production is very significant. However, most of the seedling production happened in the country deploy unimproved seed materials thus resulting in wide range of failures in several afforestation and reforestation programmes.

In the recent past, emphasis has been given to utilize improved seed materials derived from organized tree breeding and improvement programmes. This emphasis has been further strengthened due to the advancements made in tree breeding and improvement activities which necessitated the use of improved seeds collected from established seed orchards. Several seed orchards are established in wide range of species to ensure genetically improved and quality seeds with specially desired characteristics. The production of seeds in required volume for field planting is a major objective of any applied tree breeding programme. Against this backdrop, this chapter on seed orchards has been conceived and presented with the approach and methods of establishing seed orchards which will be a great utility to the tree breeders.

OBJECTIVES AND GOALS OF A SEED ORCHARD

- The main objective of the seed orchards is to produce genetically improved seeds.
- The goal of a seed orchard is to produce large quantities of seeds simultaneously maximizing genetic gain and minimizing establishment cost.
- The primary objective of seed orchard is to prevent contamination from external inferior pollen to facilitate inter and intra mating towards ensuring quality seed production.
- Above all, the purpose of seed orchards is to produce frequently, abundantly and easily harvested crops of quality seeds.

DEFINITIONS

- A seed orchard is a plantation of genetically superior trees, isolated to reduce pollination from genetically inferior outside sources and intensively managed to produce frequent, abundant and easily harvested seed crop. It is established by setting out clones (as grafting or cuttings) or seedling progeny of trees selected for desired characteristics.
- A seed orchard is a plantation of selected clones or progenies which is isolated or managed to reduce pollination from outside sources, managed to produce frequent, abundant and easily harvested crops of seed.
- Seed orchards are the most important means the tree breeder has for mass-producing seed for large improved plantations, based on the best selected trees (Faulkner, 1975).
- Seed orchards are established to mass-produce seeds of some population of which it is impossible to obtain seed in adequate quantities, without too much attention being paid to the genetic superiority of the individuals.

TYPES OF SEED ORCHARDS

Seed orchards are classified into several types based on the following two factors into considerations

- Type of reproductive material used
- Type of genetic resources deployed

Based on the type of reproductive material used, seed orchards are classified into two types

I. Seedling seed orchard

II. Clonal seed orchard

I. Seedling Seed Orchard (SSO)

Seedling seed orchards are orchards established by using seedlings obtained from genetically superior trees (Fig 10.1).

II. Clonal Seed Orchard (CSO)

These are orchards established by deploying vegetatively propagated planting materials through rooting of cutting/ grafting/ layering/ budding and in certain cases using tissue culture plants as well.

Based on the type of genetic resources deployed

i. Provenance seed orchard

ii. Progeny seed orchard

iii. Clonal seed orchard

i. Provenance seed orchard (PSO)

It is called as provenance seedling seed orchard whose components are the seedlings from diverse provenances collected from the original native habitats.

In this orchard for any species, provenances are collected from across natural ranges and evaluated through systematic provenance test. Based on the evaluation, the superior provenances are retained and the inferior provenances are rouged out and will not participate in the reproduction system. Even within the retained superior provenances, only the best individuals of best provenances are kept for seed production and other inferior individuals of a superior provenance may be eliminated. This provenance seed orchard will act as a valuable base population for any further breeding programme.

ii. Progeny seed orchard (PSO)

It is otherwise called as progeny seedling seed orchard whose components are seedlings from identified and diverse progenies. This progeny seedling orchard is similar to seedling seed orchard but with well defined progenies and act as a very valuable material for immediate adoption and incorporation in the future breeding programme

In this orchard, wide range of identified progenies are evaluated through a systematic progeny test. After evaluation, the progenies are ranked and the potential and superior progenies are retained and the inferior progenies are rouged out. Once again, the lean, lanky and inferior trees in the superior progenies are also culled to retain only the best individuals of the best progeny which will take part in reproduction process.

iii. Clonal Seed Orchard (CSO)

In this case, the orchards are established through the clonal evaluation test and after the evaluation period is over. The clones are ranked based on growth and quality attributes and only the superior and acceptable clones are retained for seed production process. The other clones are rouged out. This clonal seed orchard will help to establish rapid genetic gain and also ensure early seed production compared to other seed orchards (Fig 10.2).

There are some specialized seed orchards which are furnished below:

a. Breeding Seed Orchard (BSO)

These are orchards established and maintained for breeding purposes. The main focus of this orchard is to allow the breeding and subsequent genetic testing of greater number of parents than are normally established in a production orchard. These orchards are exclusively for breeding purposes and the management of this orchard is different from production seed orchard.

b. Extensive Seedling Seed Orchard (ESSO)

These are the orchards established with special stock from a balanced mixture of seeds from atleast 60 good parents (proven with superior combining ability) and thereafter the inferiors are culled. In ESSO, the identity of the family is maintained all through the seed collection and nursery development process to create balanced seedling lot giving equal chances for all the families.

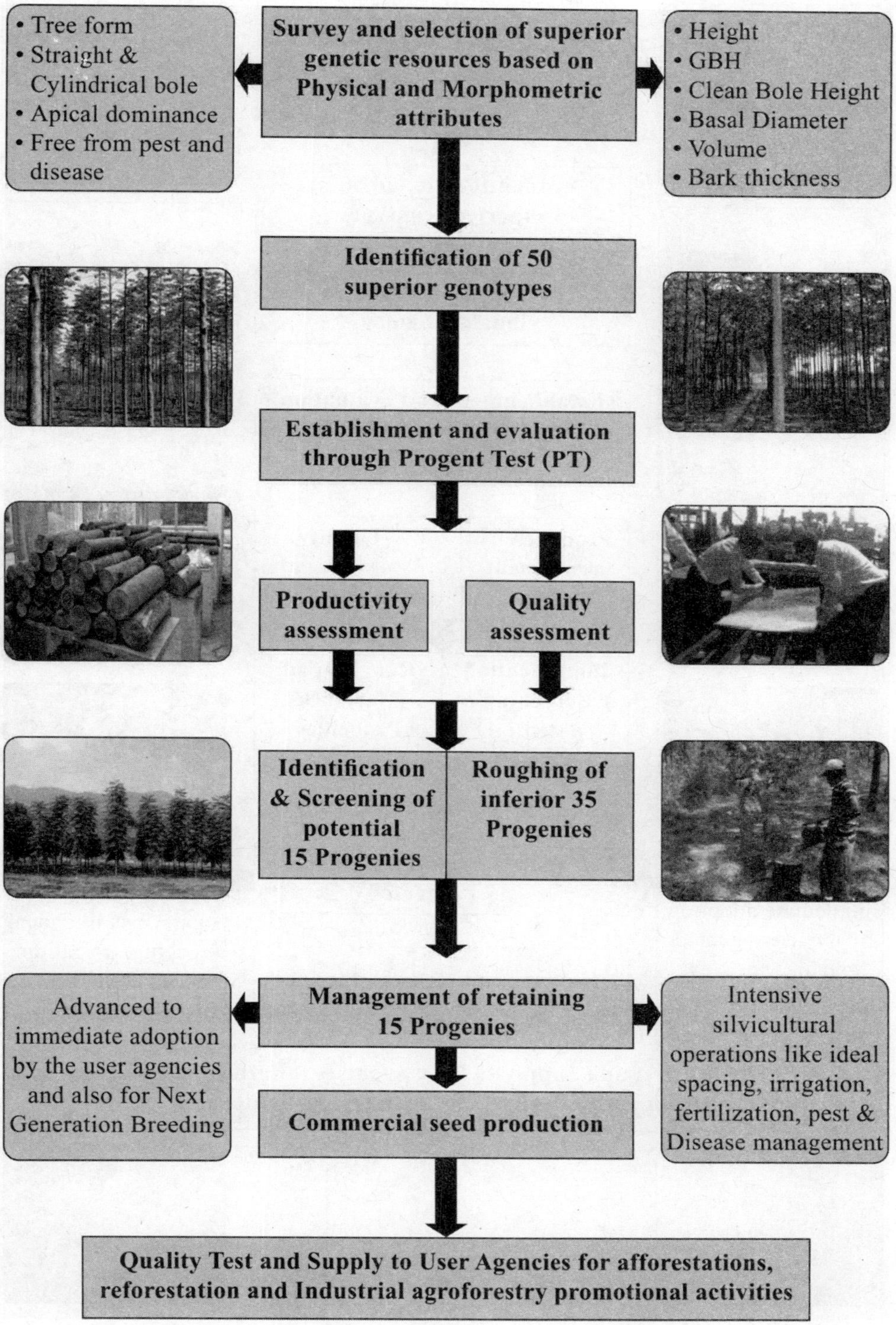

Figure 10.1. Development of Melia Seedling Seed Orchard (SSO)

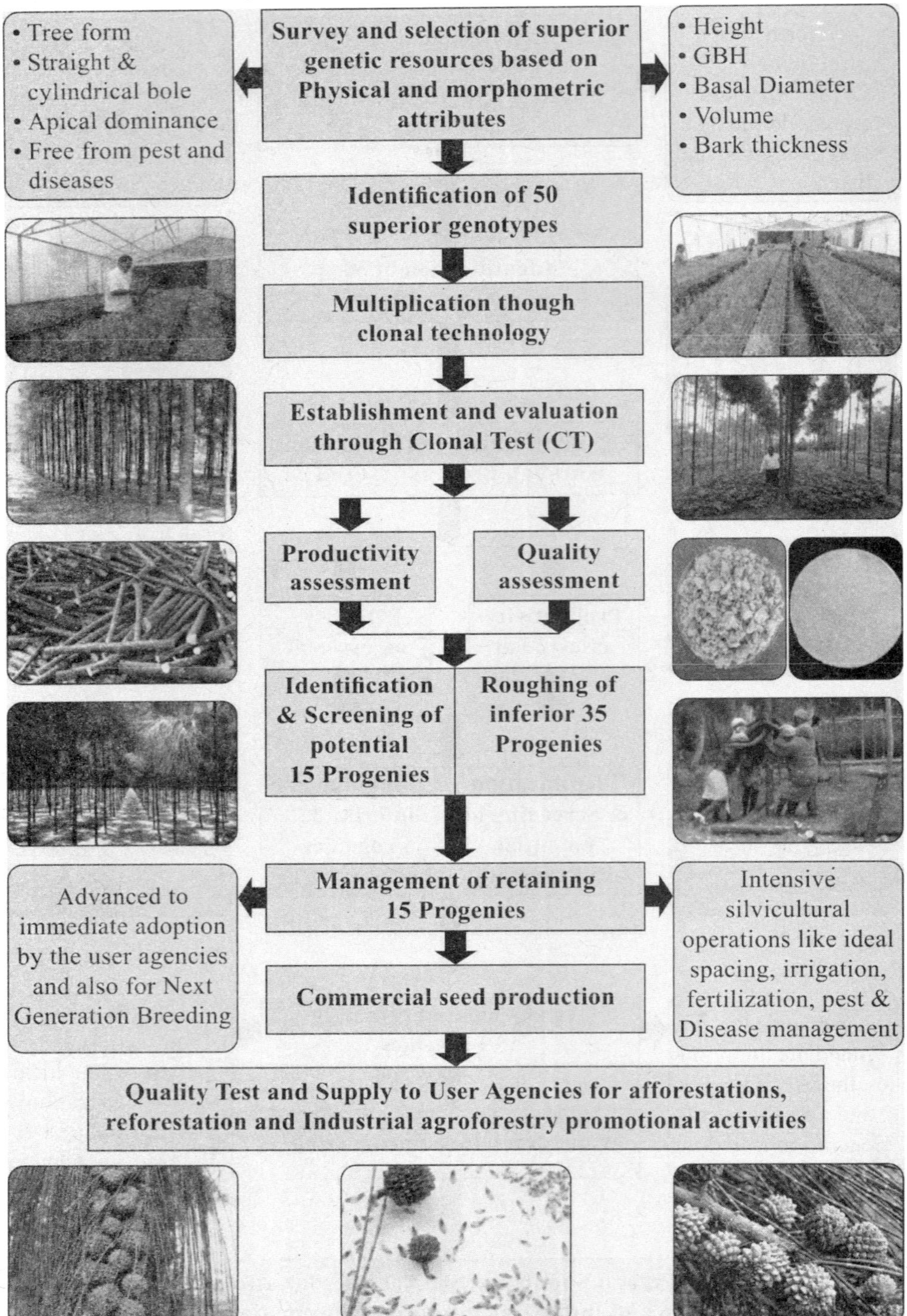

Figure 10.2. Development of Casuarina Clonal Seed Orchard (CSO)

STEPS IN SEED ORCHARD ESTABLISHMENT

i. The superior genetic stocks are to be selected from the natural stands or from plantations by using regression method or comparison tree method.

ii. From the identified trees, seeds are collected and incorporated as progenies for evaluation through progeny test. Sometimes cuttings or scions can be collected from the identified trees and propagated vegetatively for incorporation as clones and deployed for clonal test

iii. The progenies or clones are evaluated systematically both for biomass and quality. Based on the productivity and quality, superior progenies / clones are identified. The inferior families/ clones screened based on progeny / clonal test are rouged out and will not be involved in reproductive system.

iv. The retained superior families / clones are managed with suitable silvicultural practices and allowed for intermating between the families and clones for commercial production of genetically improved seeds.

v. All the activities of the seed orchard are to be maintained through systematic seed orchard register to facilitate quality and required quantity of seed production.

vi. The produced seeds are tested through standard seed testing procedures and supplied to the user agencies for establishment of industrial plantations and all afforestation and reforestation programmes.

ESTABLISHMENT OF A SEED ORCHARD

The establishment of a seed orchard is very important wherein the number of families / progenies / clones need to be decided well in advance. For any first generation seed orchards, 25-40 clones / families are generally preferred for establishment. It is also indicated that for long rotation species (40-150 years) number of families / clones may include 50 – 80 for seed production orchards. Similarly for short rotation trees (5-15 years), the population size of 10–15 clones is recommended. Once a population or family size is decided, then the size of seed orchard to be fixed. The size normally depends upon the volume of seed to be produced.

Though the size varied from species to species, it also estimated that for teak seed orchard establishment, an orchard size of 5.5 to 6.5 ha is required for planting in 50,000 ha in a span of 10 years. In this orchard, the spacing recommended is 3m × 3m and 20% culling in nursery followed by 10 % culling in transportation (Jha and Mckinley, 2002).

Similarly for semel (*Bombax ceiba*), it is suggested that the orchard size of 0.5 ha is enough to plant up to 500 ha area (Kedharnath, 1974).

Location of Seed orchard

- The Seed orchard should be located normally in good to average soil fertility conditions preferably the fallow agricultural field is good for establishment of Seed orchard.
- Seed orchard should not be located in areas with heavy wind velocity which will create pollen contamination and may lead to poor seed quality.
- The orchard should be located in places with good soil drainage in order to avoid water drainage and stagnation.
- The soils in the identified locality should have acceptable pH level in order to successfully accommodate the production population. Generally for temperate species, acidic pH level (5.5 to 6.5) and tropical tree species neutral pH (6.5 to 7.5) is preferable.
- The identified locality for Seed orchard establishment should be free from extremes of weather conditions.
- The locality should also be free from influence due to all biotic factors.

Seed Orchard Design

The design of the Seed orchard is very important towards establishment and management of the clones or the families. There are several designs practices in the world which are provided below

1. Pure Row Design
2. Chess Board Design
3. Completely Randomized Design
4. Randomized Block Design

5. Fixed Block Design
6. Rotated Block Design
7. Reversed Block Design
8. Unbalanced Incomplete Block Design
9. Balanced Incomplete Block Design
10. Cyclic Balanced Incomplete Block Design
11. Directional Cyclic Balanced Incomplete Block Design
12. Balanced Lattice Design
13. Permutated Neighbourhood Design
14. Systematic Design

All the 14 designs have advantages and disadvantages. Maciej Giertych (1977) indicated and compared the advantages and disadvantages of all the above designs which are furnished in the following table.

S.No	Particulars	Design													
	Kinds of design	1	2	3	4	5	6	7	8	9	10	11	12	13	14
1.	Avoidance of selfing in seed orchard	+	+	+	+	+	+	-	+	+	+	+	+	+	++
2.	Favouring panmixis in seed orchard			+	+	-	-	-	+	+	++	++	+	++	-
3.	Possibility of systematic thinning without altering clonal thinning	+	-	-	-	+	-	+	-	-	-				++
4.	Possibilities of orchard segment are replicates for other experiments	-	+		+	+	+	+	-	-	-	-	+	-	+
5.	Possibilities of comparing clonal performance	+	+	+	-	-	-	+	-	+	+	-	+	+	-
6.	Facilitate the relocation of clones for special cone collection for temperate trees	+	-	-	-	+	-	-	-	-	+	+	-	-	+
7.	Possibility of expansion of seed orchard	+	+	+	+	+	+	+	+	-	+	+	+	-	+
8.	Amenable for any orchard shape	+	+	+	-						-	+	-	-	+
9.	Accommodation any number of clones and ramets	+		+	+	+	+	+	-	-	-	-	-	+	+
10.	Utilization of information on Compatibility, flowering time and combining ability	+	+	-	-	+		-	-	-	-	-	-	-	+
11.	Simplicity of seed orchard design	+	+	+	+	+	+	+	-	-	-	-	-	-	+
12.	Low cost design	+	+	+	+	+	+	+	+	+	+	+	+	-	+

Note: ++ = Highly suitable; + = Suitable; - = Unsuitable

1	Pure Row Design	2	Chess Board Design
3	Completely Randomized Design	4	Randomized Block Design
5	Fixed Block Design	6	Rotated Block Design
7	Reversed Block Design	8	Unbalanced Incomplete Block Design
9	Balanced Incomplete Block Design	10	Cyclic Balanced Incomplete Block Design
11	Directional Cyclic Balanced Incomplete Block Design	12	Balanced Lattice Design
13	Permutated Neighbourhood Design	14	Systematic Design

Among these designs, the potential of Completely Randomized Design and Randomized Block Design are very well witnessed for most tree species in India due to the following reasons.

- These designs are minimising selfing among the populations
- Facilitate interacting of all genotypes in a population
- Facilitates simple and easy establishment with low cost management.
- These two designs permit the use of desired number of clones / families.

i. Completely Randomized Design

This design permits the distribution of all ramets of each clones or all members of each family randomly throughout the site.

- Number of genetically superior clones: 8
- Population density : 48 Plants /plot

CJ 1	CJ 5	CJ 3	CJ 7	CJ 2	CJ 4	CJ 6	CJ 8
CJ 4	CJ 6	CJ 1	CJ 5	CJ 8	CJ 3	CJ 7	CJ 2
CJ 2	CJ 7	CJ 6	CJ 3	CJ 5	CJ 1	CJ 8	CJ 4
CJ 7	CJ 4	CJ 2	CJ 8	CJ 6	CJ 5	CJ 3	CJ 1

ii. Randomized Block Design

This design is most preferred and applicable design wherein the total area is divided into equal blocks which is sufficient in size to house the ramets of each clone or individual of a family and the positions within each block are completely randomized.

- Number of genetically superior clones: 8
- Population density : 9 Plants /plot

MD1	MD 1	MD 1	MD 3	MD 3	MD 3	MD 7	MD 7	MD 7	MD 4	MD 4	MD 4
MD1	MD 1	MD 1	MD 3	MD 3	MD 3	MD 7	MD 7	MD 7	MD 4	MD 4	MD 4
MD1	MD 1	MD 1	MD 3	MD 3	MD 3	MD 7	MD 7	MD 7	MD 4	MD 4	MD 4
MD 5	MD 5	MD 5	MD 8	MD 8	MD 8	MD 6	MD 6	MD 6	MD 2	MD 2	MD 2
MD 5	MD 5	MD 5	MD 8	MD 8	MD 8	MD 6	MD 6	MD 6	MD 2	MD 2	MD 2
MD 5	MD 5	MD 5	MD 8	MD 8	MD 8	MD 6	MD 6	MD 6	MD 2	MD 2	MD 2

POLLEN DILUTION ZONE

Pollen dilution zone is a zone surrounding a seed orchard or seed production area in order to protect the seed orchard from contamination of outside pollen. There are, three modes of establishing pollen dilution zone, they are

i) Open type with grass or low vegetation

ii) Trees of different species, which do not pollinate with that species

iii) Large stands should be surrounded by a zone of good phenotype.

POLLEN DILUTION ZONE IN SEED ORCHARD

In seed orchards, the areas cleared between the seed stand and the possible contaminating sources is termed as pollen dilution zone. The pollen dilution zone varies with species and wind exposure. For example, for wind pollinated species, the pollen dilution zone is usually between 200 and 300 m. In case of insect pollinated species, the width of pollen dilution zone depends on the behaviour of the pollinator.

ESTABLISHMENT OF POLLEN DILUTION ZONE

Cleared areas are established between the seed stand and possible contaminating sources. The pollen dilution zone is a zone where trees of the same species and species that can hybridize with the species are removed. The wideness depends on the species and pollination pattern. The zones also serve as fire-breaks. For wind pollinated species, a short vegetation of the pollen dilution zone is desired, usually 200–400 m, depending on species and wind exposure. If there is a prevailing wind during the pollination period, the width of the pollen dilution zone may be diminished from the other sides. For insect-pollinated species, the width and the vegetation of the pollen dilution zone depends on the behaviour of the pollinator. Information should be sought on individual species.

Factors affecting the effectiveness of Pollen Dilution Zone

i) Species

ii) Wind exposure

iii) Width of pollen dilution zone

iv) Vegetation of pollen dilution zone.

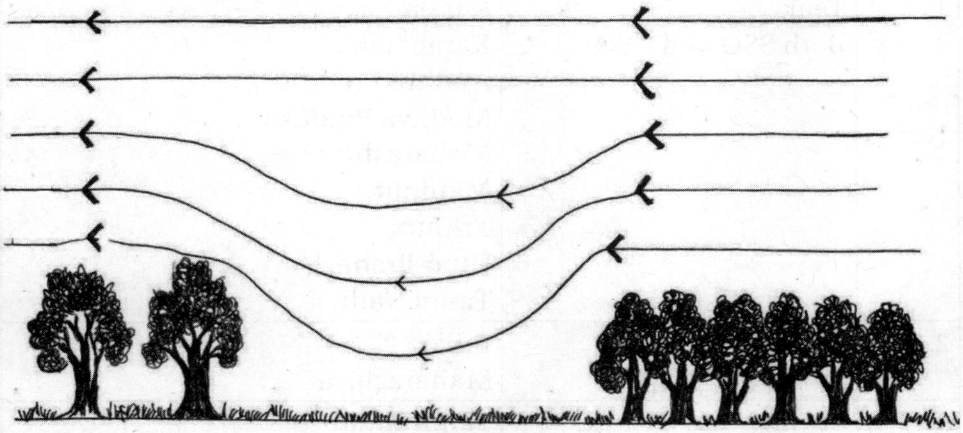

Pollen Dilution Zone

Options available in lieu of pollen dilution zone

The options available for establishment of pollen dilution zone

a) Establishment of wind break

b) Development of shelterbelts

c) Establishment of linear plantations

d) Increasing the width of the pollen dilution zone

e) Establishing plantations with sterility

SEED ORCHARDS ESTABLISHED IN INDIA

Establishment of seed orchard formed a mandatory objectives of several State Forest Departments, ICFRE & ICAR Institutes followed by Forestry Colleges of State Agricultural Universities. Several species were deployed for seed orchard establishment incorporating superior families and clones and act as a centre of seed production. The seed orchards established for major species in India is provided in the following table

S.No.	Species	State
1.	Teak Both SSO and CSO	Kerala Karanataka Andhra Pradesh Madhya Pradesh Maharashtra Manipur Tripura Uttar Pradesh Tamil Nadu
2.	*Dalbergia sissoo* SSO	Bihar Maharashtra
3.	Eucalyptus Both SSO and CSO	Tamil Nadu Uttar Pradesh
4.	*Bombax ceiba* SSO	Arunachal Pradesh Uttar Pradesh
5.	*Santalum album* Both SSO and CSO	Karnataka Tamil Nadu
6.	*Gmelina arborea* SSO	Tripura Arunachal Pradesh
7.	Casuarina Both SSO and CSO	Tamil Nadu
8.	*Pinus kesiya* SSO	Manipur
9.	*Melia dubia* SSO	Tamil Nadu

MANAGEMENT OF SEED ORCHARD

Management of seed orchard is very important to produce seeds to the maximum potential and to exploit genetic vigour of the clones or families. This is possible through efficient management of seed orchard by judicial management of all natural regeneration like soil, water and nutrients etc.

Soil Management

The texture of soil is very important which has influence on moisture, nutrient holding capacity, and compatibility, erodability and soil characteristics. Generally, the sandy loam or red loamy soils are preferred for seed orchard establishment. These soils are managed through the application of Farm Yard Manure at the rate of 12.5 MT / ha before establishment of a plantation.

Soil Fertilization

Fertilization is one of the important activities in seed orchard management in order to promote both vegetative and reproductive growth and development. Fertilization particularly, application of nitrogen for vegetative growth and phosphorous for reproductive growth is very essential in seed orchard management. The application of macro and micro nutrients may also influence the rate of production and productivity. Hence, fertilizer application should be based on soil and foliar analysis and application of soil amendments either periodically or from time to time in order to maintain vigour, growth and promote flowering. Soil pH is yet another factor which influences the vigour and growth. Generally the pH requirement of 5.5 to 6.5 is preferable for coniferous species and pH range of 6.5 to 7.5 is preferred for most broad leaved tropical species. Accordingly, the soil in the seed orchard should be managed with the application of suitable amendments to adjust and correct the pH.

Irrigation

Similar to fertilizer application, irrigation is also one of the important factors in seed orchard to maintain optimal growth and vigour. Though, installation of irrigation system is expensive in seed orchard couple of years before but now it appears to be essential investment for various species due to the water crisis. Irrigation is very essential at early stages and also during drier period preferably during summer in order to maintain in vigour and growth. Hence, suitable permanent and temporary irrigation arrangements should be made in seed orchard location.

Other silvicultural operations

Thinning/Rouging

It is always better to do thinning of bad or inferior tree in the best clones or families which will help to produce quality seeds from the best families or clones. Similarly the inferior individuals in the best family or clone are also culled to eliminate from the reproduction process and to produce seeds from best individuals of a best families or clones.

Pruning and canopy management

In seed orchard programme, pruning and branches the canopy management is very essential for several tree crops in order to induce flowering in seed orchard. The pruning and canopy management are generally acceptable methods for flower induction and management in seed orchard. Pruning particularly top pruning helps to have easy seed collection. In certain species, creation of moisture stress also found to increase flowering. In the recent past, flower inductions in seed orchards are managed for through application of gibberellins. In Brazil, flower induction in eucalyptus was initiated by the use of a growth retardant paclobutrazol. All these activities are very essential in seed orchard management practices.

Protection of seed orchard

Seed orchards are also amenable for influences due to several pests which vary from insect to diseases to animals, birds and even to human beings. Among all biological agents, pest and diseases are very serious and can be controlled by suitable systemic chemicals.

Record keeping

Record keeping is very significant in seed orchard management in order to trace the history of establishment and the associated activities carried out from establishment to current stage. The following information must be followed and recorded in seed orchard register which will help to produce seeds of significant quality and volume.

i. Date of establishment

ii. Number of clones/families

iii. Number of Ramets or individuals in a family

iv. Periodic growth attributes

v. Management of activity carried out periodical like irrigation, fertilization, thinning, pruning etc.,

vi. Age of first flowering

vii. Method of flower production

viii. Level of flower production

ix. Quantity of seed produced clone wise

x. Quantity of seed production/ha

xi. Record fertilization rate and formulation

xii. Date of application

xiii. Irrigation quantity, frequency and date

xiv. Complete weather records of the locality

xv. Occurrence of pest and disease

xvi. Treatments imposed to control pest and disease

xvii. Records of adverse environmental conditions like drought, flood, wind and rain.

xviii. Assessment of both positive and negative impacts of climatic factors on seed production.

The recording of all these factors will help the seed orchard managers towards efficient management of seed orchard to obtain abundant, frequency and easily harvested crops of seeds to satisfy the demands of various stakeholders and to establish successful afforestation and reforestation programme.

ADVANTAGES

Advantages of Seedling Seed orchard

- SSO 's are very easy to establish.
- It is applicable for those species which are recalcitrant for vegetative propagation
- SSO's will have broader genetic base than CSO
- It is possible to identify an outstanding genotype at once.

Advantages of Clonal Seed orchard

- The CSO's will help to maintain genetic identity
- The seed production in CSO's is very early and helps early seed production
- Synchronized flowering and maturity is predominant in CSO's
- The populations will have higher selection differential

- The outstanding genotypes may appear several times
- CSO's facilitates easy harvesting of seeds due to low crown branching
- Silvicultural managements like pruning and canopy management are very easy
- For species with late physiological flowering, CSO's are very good method
- Reduced selfing process results in superior seed quality

LIMITATIONS

Limitations of Seedling Seed orchard

- The seeds from SSO's may segregate which may lead to loss of genetic identity
- Selfing is possible in SSO's and may detract both quality and quantity of seed production
- SSO's will take longer period for flower and fruit production.
- The variability in height of trees in SSO's may limit the pollination mechanism
- The harvest of seed is difficult in SSO's compared to CSO's

Limitations of Clonal Seed orchard

- Development of clones needs skilled labourers which is very difficult in most cases
- Development of clones involves huge cost and expenditure
- It will also take longer time to develop a clone
- The genetic base is very narrow
- CSO's are susceptible to wind and other abiotic stresses very easily.

SUMMARY

Production and supply of genetically improved and quality seeds have a major impact on productivity and profitability of industrial wood plantations and success of several afforestation and reforestation programme. This quality seed production is possible through the establishment of production seed orchard which may be established either through seedling based seed orchard or through clone based seed orchard. These seed orchards should be established in a very designed manner and managed with suitable silvicultural operations in order to produce quality seeds. The requirement of suitable environmental conditions like soil and water coupled with management options like irrigation, fertilization, thinning, pruning, flower induction, protection from all pests etc., are very critical towards successful establishment and management of seed orchards. All these activities need to be recorded systematically which will act as a history of events not only for inspection but also help the seed orchard manager to efficiently manage the seed orchard in order to produce quality seeds in a required volume. This chapter details all these activities of seed orchard establishment and management which will be of great use to breeders and seed production industries and could play a vital role in plantation forestry development programme.

11

FOREST GENETIC RESOURCES

INTRODUCTION

Forest Genetic Resources are the biological resources which represent the variety and variability that exist at the levels of genetic, species and population levels. These genetic resources are under intensive pressure due to population growth, deforestation, habitat alteration and indiscriminate extraction of forest products. This besides, the industrialization, urbanization, development and communication, development, conversion of forest land for non forestry purposes like river valley projects, dams, electric line establishment and other communication system have increased the rate of deforestation and threatened the wide range of biological resources. To resolve all these issues, the conservation of forest genetic resources has gained significant importance. Several actions and policies that ensure the continued existence, evolution, reproduction and their availability not only for current generation but also for future generation have been taken by the government both at state and central levels. This chapter elaborates the basic concepts of biological resources, genetic resources of various utility based forest resources, conservation methods and the associated legal, policy and conventions.

DEFINITION OF TERMINOLOGIES

Forest Genetic Resources :
- Forest Genetic Resources are the resources which include the genetic variation in trees of potential or present benefit to the human beings (FAO, 1989).
- This forest genetic resources is also defined individually as forest, as genetic and also as resources.

- Forest Genetic Resource is defined as the heritable materials maintained within and among tree and other woody plants species that are of actual or potential economic, environmental, scientific or social values.

Forest : It denotes a stand, population or landscape of trees which includes other associated woody plants.

Genetic : It refers to variation of genetic origin and variation of genes at different levels which include the following

1) Variation between species (Species Variation)
2) Variation between population within species (Population Variation)
3) Variation between individual trees within population (Tree variation)

Resources : It refers to the use of genetic variation which is considered to be of potential value to the human both at present and in the near future

FLORISTIC REGIONS OF INDIA

India represents 11 Floristic regions which are furnished below:

1. North west Himalaya
2. Indo-Gangetic Plains
3. Eastern Himalayas (Arunachal Pradesh and North Bengal)
4. Assam (North eastern India excluding Arunachal Pradesh and North Bengal)
5. Central India
6. Arid zone
7. Northern Western Ghats and Northern west coast and Lakshadweep
8. Southern Western Ghats and Southern west coast and Lakshadweep
9. Deccan Plateau
10. Eastern Ghats and Coromandel coast
11. Andaman and Nicobar Islands

BIOGEOGRAPHIC ZONES OF INDIA

India is home to large diversity of species and ranks six among the twelve mega biodiversity centres of the world. India is a part of four terrestrial biodiversity hotspots which are as follows

1) Himalayan region
2) Indo Burma
3) Western Ghats
4) Sri Lanka and Sundalands

India ranks tenth in the world for plant biodiversity accounts for 11% of the floral diversity of the world. This biologically rich diverse country is represented by two realms.

NAME OF THE REALM	REGION
Paleo- artic	Himalayan Region
Malayan Realm	Rest of the country

India is also housed with 10 biogeographical zones which included 27 biogeographical provinces (MOEF, 2009).

S.No	BIOGEOGRAPHIC ZONES	BIOGEOGRAPHICAL PROVINCES
1.	Trans Himalaya	1A: Himalaya - Ladakh Mountains 1B: Himalaya -Tibetan Plateau 1C: Trans - Himalaya Sikkim
2.	The Himalaya	2A: Himalaya - North West Himalaya 2B: Himalaya - West Himalaya 2C: Himalaya - Central Himalaya 2D: Himalaya - East Himalaya
3.	The Indian Desert	3A: Desert – Thar 3B: Desert – Katchchh
4.	The Semi Arid	4A: Semi - Arid - Punjab Plains 4B: Semi - Arid – Gujarat, Rajputana
5.	The Western Ghats	5A: Western Ghats - Malabar Plains 5B: Western Ghats -Western Ghats Mountains
6.	The Deccan Peninsula	6A: Deccan Peninsular - Central Highlands

S.No	BIOGEOGRAPHIC ZONES	BIOGEOGRAPHICAL PROVINCES
		6B: Deccan Peninsular - Chotta Nagpur 6C: Deccan Peninsular - Eastern Highlands 6D: Deccan Peninsular - Central Plateau 6E: Deccan Peninsular - Deccan South
7.	The Gangetic Plains	7A: Gangetic Plain - Upper Gangetic Plains 7B: Gangetic Plain - Lower Gangetic Plains
8.	The Coasts	8A: Coasts - West Coast 8B: Coasts - East Coast 8C: Coasts – Lakshdweep
9.	Northeast India	9A: North - East - Brahamputra Valley 9B: North - East – North East Hills
10.	Islands	10A: Islands – Andamans 10B: Islands – Nicobars

CHARACTERISTICS OF FOREST GENETIC RESOURCES

- Forest Genetic Resources are long lived and develop their own natural mechanism to maintain high levels of genetic variation within the species.
- Most Forest Genetic Resources have high rates of out crossing and have the capacity to disperse the pollen on the long distance.
- Most Forest Genetic Resources have several dispersal mechanisms which help to transfer the seed to various distances and help to spread the biological diversity.
- The natural out crossing mechanism coupled with the native environment enable forest tree species to evolve the most genetically diverse organisms.
- The chromosomes and DNAs of Forest Genetic Resources are generally large and seemingly inexplicable differences and

variations in chromosome number, ploidy levels and genome size both among trees and between trees and other organisms

- Whole genome duplication which is otherwise polyploidization is also one of the characteristics features of most Forest Genetic Resources which lead to drastic increase in species richness.
- Some of the species in Forest Genetic Resources have longevity which results in persistence over millions of years as evidenced by *Ginkgo biloba* and *Wollemia nobilis* (the species evolved 61 million years ago).
- Forest Genetic Resources are larger in sizes which include the biggest and tallest organisms in the plant.

  Example : *Sequoia sempervirens* (150 m tall and 530 tonnes of weight)

  Eucalyptus regnans (100 m height)
- Another important characteristics of Forest Genetic Resources is that, it exhibits diverse breeding and reproductive system due to the presence of monoecious, dioecious, hermaphroditic and polygamous.
- The Forest Genetic Resources exhibit higher level of species diversity which included approximately 8000 to 100000 tree species and have been described and currently accepted as valid and unique species diversity. Similarly Forest Genetic Resources have characteristics features of within species diversity (intraspecific diversity and expressed at various levels).

IMPORTANCE OF FOREST GENETIC RESOURCES

- Forest Genetic Resources are an excellent biological resources and provide wide range of direct and indirect benefits to the society and played a significant role in improving the socio-economic status of the people.
- More than 25% of the world's population (1.6 billion people) rely on the forest resources for their livelihood.
- The Forest Genetic Resources play a vital role providing the raw material resources of various wood based industries like timber, pulp and paper and biomass industries and contributes significantly towards the trade of various forest products.

Similarly, the Forest Genetic Resources along with agriculture and horticulture contribute significantly for the production and trade.

- The Forest Genetic Resources have a very prominent role in meeting food, fuel, fibre and fodder needs of the local people and the associated livestock development.
- Forest Genetic Resources are rich source of medicinal value and contribute significantly due to the growth and development of medicinal systems like Siddha and ayurvedic system. The World Bank (2002) estimated that globally 1 billion populations depend on drugs derived from forest plants for their medicinal needs.
- The Forest Genetic Resources and the associated economic values of several wood based industries are contributed towards GDP (Gross Domestic Product). Several products developed through trees of significant Forest Genetic Resources and amenable for export and play a vital role in foreign exchange contribution.
- The trade of global forest products has been estimated at $US 255 million during 2011 which included 40 % contribution to the developing countries and act as a major source of a revenue to these countries. It is also estimated that the forest products industry is able to generate 49 million job opportunities and could act as a sustained income and employment generation enterprise.
- The Forest Genetic Resources provided wide range of Non Timber Forest Products (NTFPs) and provide income earning opportunities for the rural women, thus contribute to women empowerment and gender equality.
- Forest Genetic Resources have also played an extremely important role in supporting agricultural production through agroforestry systems by providing shelter, shade, and fertility improvement and reduce the erosion and mitigating the negative impacts of floods.
- The Forest Genetic Resources in general and some trees in particular have played important role in alleviating poverty during the times of hardship and crop failure.

- The genetic diversity that exists in tree species contributes significantly towards economic utility in forest plantations and agroforestry system by increasing the output of superior forest products with low levels of inputs.
- The Forest Genetic Resources provide wide variety of environmental services like water augmentation, carbon sequestration, soil fertility amelioration and maintenance of ecological and ecosystem processes.
- The ecosystem services of Forest Genetic Resources are gaining attraction across the world due to their role in extending provisional, regulation, cultural and supportive services.
- Above all, the role of Forest Genetic Resources in mitigating and adopting to growing climate change associated issues is very important to resolve the risks and uncertainty.
- Above all, the rich biological diversity of Forest Genetic Resources supports wide range of species and maintains the ecosystem richness.

MAJOR AND POTENTIAL FOREST GENETIC RESOURCES OF INDIA

i. Timber Genetic Resources

Timber plays a vital role in meeting the wood demands of both domestic and industrial utility. Timber has profound applications in construction industry, defence, Railways, sports goods, furniture and agricultural implements. It is estimated that the country would need more than 65 mill m^3 of wood from long rotation (LR) species in order to meet the raw material requirement of the industries indicated above. India was once a major exporting country for timber which has now become major importer of timber particularly the species like teak. The legal restriction on felling in the natural forest has promoted agroforestry particularly timber based agroforestry. Hence systematic tree improvement programme has been initiated by various institutes in order to improve the productivity of the species. The major timber species and the associated improved genetic resources are furnished below.

S. No	Common name	Botanical name	Family	Other species
1.	Rose wood	*Dalbergia latifolia*	Fabaceae	*Dalbergia sissoo* *Dalbergia melanoxylon* *Dalbergia maritime*
2.	Teak	*Tectona grandis*	Lamiaceae	*Tectona hamiltoniana* *Tectona philippinensis*
3.	Cedar	*Cedrus deodara*	Pinaceae	*Cedrus brevifolia* *Cedrus atlantica* *Cedrus libani*
4.	Babul	*Acacia nilotica*	Leguminaceae	*Acacia auriculiformis* *Acacia planiferans* *Acacia leucocephala* *Acacia ferruginea* *Acacia catechu* *Acacia mearnsii* *Acacia holosericia*
5.	Neem	*Azadirachta indica*	Meliaceae	
6.	Mahogany	*Swietenia macrophylla*	Meliaceae	*Swietenia mahogany*

Status of Timber genetic resources at FCRI, TNAU

S. No	Common name	Name of the Genus/ Species	Seed source	Progeny	Potential Clone / Progeny / Seed source / Variety
1.	Teak	*Tectona grandis*	30		❖ Syyaburry - MTPTK07 ❖ Nilambur - MTPTK21 ❖ Chandrapur - MTPTK16
2.	Portia tree	*Thespesia populnea*		30	
3.	Mahogany	*Swietenia macrophylla*		24	❖ FCRISM19 ❖ FCRISM15 ❖ FCRISM13
4.	African Mahogany	*Khaya senagalensis*		2	❖ FCRIKS18

ii. Pulpwood Genetic Resources

India is the robust producer and consumer of paper and paper products which accelerated the demand for pulpwood raw material. The country is housed with over 700 paper industries and among them 30-40 % of the industries are wood based industries. The growing population, industrialization, changes in education policy and science and technology development have increased the demand for paper and paper products. The per capita paper consumption has increased to 13 kg / annum from 8-10 kg/annum. This has attracted promotion of industrial agroforestry plantation to cater to the needs of pulp and paper industries. This increasing attention has also equally attracted development of systematic tree breeding and improvement programme in various species like casuarina, eucalyptus, populus, melia and kadam as a source of pulpwood genetic resources. The institutes like TNAU, ICFRE and industries like ITC, WIMCO have developed wide range of clones in several pulpwood species and are deployed in operational pulpwood developmental programme. The potential pulpwood genetic resources currently used in pulp and paper production are furnished below.

S. No	Common name	Species	Family	Other species
1.	*Eucalyptus*	*Eucalyptus tereticornis*	Myrtaceae	*Eucalyptus camaldulensis* *Eucalyptus citriodora* *Eucalyptus globulus* *Eucalyptus grandis*
2.	*Casuarina*	*Casuarina equisetifolia*	Casuarinaceae	*Casuarina junghuhniana* *Casuarina cunninghamiana*
3.	Subabul	*Leucaena leucocephala*	Leguminaceae	*Leucaena diversifolia*
4.	Acacia	*Acacia hybrids*	Leguminaceae	
5.	Bamboo	*Bambusa bamboo*	Poaceae	[*B.balcooa, B. bambos, B. tulda, B. vulgaris* (Green), *B. vulgaris* (Yellow), *D. strictus*]

Status of Pulpwood genetic resources in FCRI, TNAU

S. No	Common name	Name of the Genus/Species	Seed source	Progeny	Clone	Hybrid clone	Potential FGR
1.	Eucalyptus	*E. tereticornis, E. camaldulensis, E. toreliana, E. alba, E. polycarpa, E. microtheca, E. tasselaris, E. urophylla, E. urograndis*	65		112	35	❖ Eucalyptus MTP 1
2	Casuarina	*C.equisetifolia, C. junghuhniana, C.cunninghamiana*	12	30	66	2	❖ Casuarina MTP 1 ❖ Casuarina MTP 2
3.	Subabul	*Leucaena leucocephala*		30			❖ FCRI LL 15
4.	Lannea	*Lannea coromendalica*		30			
5.	Shisam	*Dalbergia sissoo*			35		❖ MTPDS 18
6.	Acacia	*Acacia hybrids*				9	❖ MPM-AH 35
7.	Bamboos	[*B.balcooa, B. bambos, B.tulda, B. vulgaris* (Green), *B. vulgaris* (Yellow), *D. strictus*]		15	55		
8.	Melia	*Melia dubia* *Melia composita* *Melia volkensii*	20	50	52	-	❖ Melia MTP 1 ❖ Melia MTP 2 ❖ MD-26 ❖ MD-12 ❖ MD-44

iii. Plywood Genetic Resources

The growing demands for wood and wood products coupled with declining supply from natural forests have attracted use of engineered wood products like plywood as an alternate to solid wood. The changes in housing policy, growing attraction towards interior decorations and reduced timber supply have increased the demand for composite or engineered wood products. India is housed with over 2000 organized plywood industries for which the wood requirement has estimated at more than 15 million m^3. This besides, wide range of unorganized veneer industries are established in the country and the raw material demand for these sectors is growing at an alarming rate. Hence organized agroforestry has been promoted in northern India deploying the superior genetic resources of poplars and eucalyptus and successful productivity and profitability improvement has been witnessed. Similarly in Tamil Nadu, TNAU has identified wide range of alternate plywood genetic resources and within them several HYSR clones have been developed and deployed in plywood based agroforestry development programme in association with major plywood industries in the country. The major genetic resources amenable for plywood production are furnished below.

S. No	Common name	Species	Family	Other species
1.	Poplar	*Populus deltoides*	Salicaceae	*Populus nigra* *Populus tremula* *Populus gamblei* *Populus simonii*
2.	Eucalyptus	*Eucalyptus*	Myrtaceae	*Eucalyptus tereticornis,* *Eucalyptus camaldulensis,* *Eucalyptus toreliana,* *Eucalyptus alba,* *Eucalyptus polycarpa,* *Eucalyptus microtheca,* *Eucalyptus tasselaris,* *Eucalyptus urophylla,* *Eucalyptus urograndi*
3.	Melia	*Melia dubia*	Meliaceae	*Melia azedarach* *Melia tomentosa* *Melia volkensii*

Status of Plywood genetic resources at FCRI, TNAU

S.No	Common name	Name of the Genus/Species	Seed source	Progeny	Clone	Potential FGR
1.	Melia	*(M. dubia, M.azedarach, M.composita)*	20	50	52	❖ Melia MTP 1 ❖ Melia MTP 2 ❖ Melia CW 10 ❖ MD-44 ❖ MD-7 ❖ MD-19
2.	Poplar	*Populus deltoides*			38	
3.	Kadam	*Neolamarckia cadamba*	20	50	25	❖ Kadam MTP 1
4.	Toon	*Toona ciliata*	5	30	1	❖ TC 02
5.	Umberlla tree	*Maesopsis eminii*		15		❖ VAL-ME 7
6.	Acrocarpus	*Acrocarpus fraxinifolius*	20	30		❖ FCRIAF28
7.	Chukrasia	*Chukrasia tabularis*	15			❖ FCRI CT SS
8.	Eucalyptus	Eucalyptus			15	❖ Eucalyptus LBT 01

iv. Matchwood Genetic Resources

India is one of the leaders in match splints and box production and 80 % of the matches produced in the country are from the states of Kerala and Tamil Nadu. These industries predominantly used only *Ailanthus excelsa* and *Ailanthus triphysa*. Recently these industries started deploying popular imported from European countries. Hence TNAU has initiated systematic tree improvement programme in *Ailanthus excelsa* and also identified alternate genetic resources for use in match industries. The potential genetic resources of *Ailanthus excelsa* and alternate match wood genetic resources are presented.

S. No	Common name	Species	Family	Other species
1.	Tree of heaven	*Ailanthus excelsa*	Simaroubaceae	*Ailanthus triphysa* *Ailanthus malabarica* *Ailanthus grandis*
2.	Kadam	*Neolamarckia cadamba*	Rubiaceae	*Neolamarckia cadamba*

Status of Matchwood genetic resources in FCRI, TNAU

S. No	Common name	Name of the Genus/Species	Seed source	Progeny	Potential FGR
1.	Tree of heaven	*Ailanthus excelsa*		30	❖ FCRI AE 6 ❖ FCRI AE 16 ❖ FCRI AE 26
2.	Halmaddi	*Ailanthus triphysa*	1		❖ FCRI AT 01
3.	Kadam	*Neolamarckia cadamba*	20	50	❖ Kadam MTP 1

v. Endangered Genetic Resources

The forests in the country have been degraded after independence due to anthropogenic pressure, diversion of forest land for agriculture and allied activities, river valley projects industrialization and encroachment which have taken heavy toll of Indian forest and the associated forest genetic resources. Several endemic species have been degraded at an alarming rate and are translated to a stage of endangered category. Among several species, the potential of Sandal wood (*Santalum album*) and

Redsander (*Pterocarpus santalinus*) are very well witnessed. This has attracted organized improvement programme towards conservation and rehabilitation of these two genetic resources and the status of potential FGR are provided below.

S. No	Common name	Species	Family	Other species
1.	East Indian Sandal Wood	*Santalum album*	Santalaceae	*Santalum paniculatum* *Santalum yasi* *S. obtusifolium* *S. lanceolatum* *S. fernandezianum* *S. spicatum*
2.	Redsander	*Pterocarpus santalinus*	Fabaceae	*Pterocarpus amazonum* *Pterocarpus dalbergioides* *Pterocarpus indicus* *Pterocarpus macrocarpus* *Pterocarpus marsupium* *Pterocarpus officinalis* *Pterocarpus santalinoides*

Status of Endangered Genetic Resources at FCRI, TNAU

S. No	Common name	Name of the Genus/Species	Progeny	Potential FGR
1.	East Indian Sandal Wood	*Santalum album*	30	❖ FCRI SA 16 (Tirupattur)
2.	Redsander	*Pterocarpus santalinus*	24 + 30	❖ TNRS-01 ❖ FCRIRS 07

vi. Biofuel Genetic Resources

Energy is the fundamental requirement for all developmental process and India is one of the energy dependent countries. India has 0.5 % of the oil and gas resources of the world but housed with 16 % of the world's population which results in heavy dependence on the imported oils. More than 80 % of the country's energy demand is met from imports predominantly for transport, industry and agriculture sector. The increasing use of fossil fuels also act as a major contributor of increasing pollution in the country and the associated climate change related issues. This has necessiated promotion of various renewable energy resources and among them the potential of forest based biofuel is very well witnessed. Indian forest supports over 100 different Forest Genetic Resources which are suitable for biofuel production and among them only few species are improved through systematic tree breeding approaches. The various species and the associated improved genetic resources are furnished below.

S. No	Common name	Species	Family	Other species
1.	Jatropha	*Jatropha curcas*	Euphorbiaceae	*Jatropha intergerrima* *Jatropha gossypifolia* *Jatropha maheswari* *Jatropha glandulifera* *Jatropha multifida* *Jatropha tanjorensis* *Jatropha villosa var. villosa* *Jatropha podagrica*
2.	Pungam	*Pongamia pinnata*	Fabaceae	
3.	Mahua	*Madhuca longifolia*	Sapotaceae	*Madhuca latifolia* *Madhuca butyracea* *Madhuca neriifolia* *Madhuca bourdillonii*
4.	Neem	*Azadirachta indica*	Meliaceae	

Status of Biofuel genetic resources at FCRI, TNAU

S. No	Common name	Name of the Genus/Species	Seed source	Progeny	Clone	Hybrid clone	Potential FGR
1.	Jatropha	*Jatropha curcas* and back cross derivatives	35	106	50	35	❖ CJH 5 ❖ CJH 9 ❖ CJH 10 ❖ CJH 12 ❖ CJH 13 ❖ HC 21 ❖ HC 34
2.	Karanja	*Pongamia pinnata*		66	25		
3.	Mahua	*Madhuca latifolia*		33	5		
4.	Paradise Tree	*Simarouba glauca*	20	28	5		❖ MTP 1
5.	Undi	*Calophyllum inophyllum*		15	1		❖ FCRICI 14

vii. Coniferous Genetic Resources

India is a major consumers of wood and wood products obtained both from soft woods and hard woods. Among these two categories, the potential of soft woods particularly the wood obtain from various *Pinus* spp are very well documented. The wood from pines have wide range of utilities like pulp and paper, packing cases, specialty ply woods and also as a specialty timber in beach house construction industry. Some of the tropical pines like chir pine and blue pine have very well received improvement strategies and the genetic resources associated with these two species are presented below.

S. No	Common name	Species	Family	Other species
1.	*Pinus*	*Pinus roxburghii*	Pinaceae	*Pinus wallichiana* *Pinus gerardiana* *Pinus kesiya* *Pinus oocarpa* *Pinus radiata*

viii. Mangrove Genetic Resources

Mangroves are unique and one of the distinct ecosystems housed with wide range of species. It is an ecological group of halophytic plants and includes variety of complex plant communities. India is housed with almost 5 % of the world's mangrove vegetation and accounts for nearly 4921 sq.km of the total designated forest area. The mangrove ecosystem are recognized as most productive and diverse ecosystem comprises the dominant salt tolerant, sclerophyllous broadleaved trees along with ferns, orchids, lichens, non mangrove halophytes, sea grasses and seaweeds. The major mangroves and the associated dominant species distributed are furnished below.

S.No	Mangroves	Number of species present	Dominant species
I. East Coast of India			
i.	Sunderbans	26	*Excoecaria agallacha* *Ceriops decandra* *Sonneratia apetala*
ii.	Mahanadi	26	*Excoecaria agallacha* *Ceriops decandra* *Sonneratia apetala*

S.No	Mangroves	Number of species present	Dominant species
iii.	Godavari	17	*Excoecaria agallacha Avicennia officinalis Sonneratia apetala*
iv.	Krishna	12	*Avicennia marina*
v.	Pichavaram and Muthupet	11	*Avicennia marina*
II. West Coast of India			
i.	Gulf of India	8	*Avicennia marina*
ii.	Gulf of Khambat	8	*Avicennia marina*
III. Andaman and Nicobar Islands			
i	Andaman and Nicobar Islands	24	*Rhizophora apiculata Rhizophora mucronata Ceriops tagal*

(*Source:* Selvam, 2003)

ix. Sacred groves Genetic Resources

Scared groves are traditionally called as temple forests which are one of the oldest forms of conserving natural forest. These sacred groves are predominantly community linked forest conservation approach and India has very long tradition of nature conservation in resource rich forested landscapes of Western Ghats and North Eastern parts of the country. The characteristic feature and the genetic resources associated with sacred groves are presented.

Characteristics of Sacred Groves

- The general floristic composition and physiognomy of vegetation of the sacred groves are typically like the low level evergreen forest.
- The vegetation in undisturbed groves is luxuriant and comprises several stories of trees mixed with shrubs, lianes and herbs.

- The ground is humus laden and covered with litter.
- The macro-fungus are abundant coupled with the populations of ferns.
- The water bodies in and around the sacred groves attracts the algae and water plants which grow very gregariously.
- Floristic variations have occurred in many sacred groves exposed to human and animal interferences and climatic and edaphic changes.
- Some characteristic morphological adaptations accompany the members of the sacred groves. Broad prominent buttresses are seen in most of the trees like *Ficus* spp, *Hopea parviflora* etc. which attain enormous heights with tall bole, the branches confining to the top most region

Composition of Sacred Grove Genetic resources

The composition of sacred grove involve tall trees in the top canopy followed by co dominance in second storey and small trees herbs and shrubs in the ground layer.

a. Tropical Sacred Groves – Kerala

Top canopy is represented by species like *Artocarpus hirsutus, Vateria indica, Hopea parviflora, Hopea ponga, Alstonia scholaris* etc.

Second storey consists of *Macaranga peltata, Mimusops elengi, Hydnocarpus wightiana, Holigarna arnottiana* etc. Common lianas are *Strychnos colubrina, Anamirta cocculus, Tetracera akara, Acacia intsia* etc. *Chassalia curviflora, lxora nigricans, Ixora brachiata* are constituents of the shrubby layer.

The ground layer is usually formed of seasonal members such as, *Centella asiatica, Aerva lanata, Andrographis paniculata, Biophytum sensitivum*.

Common climbers are Calamus spp, *Anamirta* cocculus, *Strychnos colubrina, Calycopteris floribunda* etc. Other climbers commonly found in these groves belong to Vitaceae, Menispermaceae, Asclepiadaceae and Apocynaceae families.

Total parasites like *Cassytha filiformis* and semiparasites like *Loranthus* spp, are also common. Common constituents of the shrubby layer include species like *Memecylon umbellatum, Pavetta indica, Chassalia ophioxyloides, lxora* spp etc.

The ground layer is usually thickly populated with species which prefer humus and love shade. Along with a few angiosperms, ferns, Selaginellas and many macrofungi like species of Agaricus, etc also occur.

Dead trunks of fallen trees harbor a variety of Polyporales especially species of Fomes and Polyporus are common as evidenced in Iringole Kavu.

b. Evergreen sacred groves

Typical evergreen elements of the tree layer of these sacred groves are *Artocarpus hirsutus, Hopea parviflora, Mesua ferrea, Vateria indica, Ficus* spp etc. There are also a few semi evergreen representatives like *Fagraea ceylanica, Murraya exotica, Samadera indica* etc. A few deciduous and semi deciduous members also are seen. Semideciduous representatives are species of Cinnamomum, *Syzygium cumiini, Litsea* spp, *Mangifera indica* etc. Species like *Mallotus philippensis, Dillenia pentagyna and Trema orientalis* are some of the deciduous species.

Semi-evergreen representatives are also found among the undergrowth vegetation. The common members being *Pavetta indica, Curcuma aromatica, Nephrolepis spp* etc. Epiphytic orchids like *Vanda spathulata, Vanda tessellata, and Bulbophyllum spp* are also common.

c. Mangrove sacred groves

A very interesting constituent of the sacred groves is represented by members of the typical mangroves swamp found in poorly drained regions of South Travancore with a very long rainy season. Common members represented are *Myristica magnifica* with prominent dense stilt roots, *Myristica malabarica. Hydnocarpus* spp. *Eugenia* spp. etc. These members are known to develop a high profile of humidity in the surroundings. Their presence in certain groves is justified as they promote luscious growth of undergrowth mostly members of Araceae, Urticaceae. Zingiberaceae, Acanthaceae etc. The exact physiological implication behind this high humidity is not experimentally proved, though it may be ascribed to very high transpiration rates of the leaves of these trees.

FACTORS AFFECTING FOREST GENETIC RESOURCES

Forest Genetic Resources are the heritable materials that are maintained within and among trees and other woody plants which are of actual or potential, economic, environmental, scientific and societal values. These Forest Genetic Resources are playing a vital role in providing provisional, regulation, cultural and supportive services to the society and play a critical role in adaptation and evolutionary process. However due to the population growth, increasing urbanization, industrialization and the associated technology development have demanded more energy and wood products both for industrial and domestic utility.

The current demand is expected to increase 40 % in next two decades which are the major drivers of changes in ecosystem and threat to Forest Genetic Resources. The major consequences of land use changes are the increased population pressure which results in conversion of forest land to agricultural, which is a major factor for the loss of genetic resources. This besides, over harvesting, degeneration due to extreme climatic events, forest fire, grazing, deforestation and illicit felling are the major drivers, thus resulting in loss of wide range of Forest Genetic Resources. These activities also result in extinction of few genetic resources. FAO (2014) has identified the major factors that are responsible for loss of Forest Genetic Resources are provided below

i. Conversion of forest for agriculture and allied activities
ii. Conversion of forest for industrial plantations
iii. Increasing demand for energy
iv. Unsustainable harvest and utilization
v. Growing cattle population
vi. Climate change
vii. Climate impact on species and ecosystem
viii. Epidemic status of pest and diseases
ix. Change in the fire regimes
x. Invasive species
xi. Genetic pollution

i. **Conversion of forest for agriculture and allied activities**

The growing global demand for land through the production of agricultural commodity is a major threat to loss of forest and the associated Forest Genetic Resources. During the period between 2010 and 2011, the permanent crop areas have increased by 15.7 % globally and in the least development country the increase was upto 22 %.

ii. **Conversion of forest for industrial plantations**

The growing demand for wood and wood products both within the country and outside along with decline in the supply of wood from natural forest have necessitated promotion of large scale plantation developmental programme by replacing natural forest towards production of timber, wood pulp, oil palm and other related forest products. These plantation are covering an area of over 109 million ha worldwide which predominantly replaced the natural place.

iii. **Increasing demand for energy**

Firewood and charcoal forms a major portion of domestic consumption with developing countries including India which is one of the main causes of forest degradation in these countries. The growing population coupled with increasing demands for both domestic and industrial requirements are also playing a vital role in destruction of forest resources. The growing concerns on energy policies and the increasing attraction towards the use of biomass based renewable energy both in developed and developing countries may have severe implications on forest and forestry resources.

iv. **Unsustainable harvest and utilization**

Increasing unsustainable harvest of forest due to over harvesting and over exploitation are the major factors causing serious threat to Forest Genetic Resources. This unsustainable harvest and utilization can seriously erode genetic diversity which may take long period for rehabilitation of these species. The species which are distributed rarely may become extinct due to over harvesting like that of *Thuja sutchuenensis,* in China, *Santalum album, Pterocarpus santalinus* and agar wood in India. Similarly, over harvesting and exploitation of NTFPs in several forests has caused serious threat to Forest Genetic Resources.

v. Growing cattle population

The growing livestock population is a serious threat to many fodder trees and can have a destructive impact on the genetic resources. Forest resources are the main source of fodder particularly in lean season for the grazing animals located near by the forest area. The browsing animals especially goats will have a serious impact on the tree vegetation and rapid removal of recruits are very well witnessed due to overgrazing.

vi. Climate change

Climate change due to industrial revolution and the associated increasing levels in the atmosphere are of great concern for Forest Genetic Resources. The destruction of forest, burning of fossil fuel, growing industries, increasing use of vehicles etc., accounts for nearly 20% of the green house gas emission and increase the level of carbon-di-oxide in the atmosphere. The climate alteration and increased occurrence of extreme climate events are pronounced as a threat to Forest Genetic Resources by several countries. Increasing mortality, delayed flowering, migration of pollinators, erratic temperature factor etc., may play a drastic role not only in loss of Forest Genetic Resources but their migration and further extinction. The impact of climate change is more vulnerable in tropical forests where in a small change in climate is likely to affect the time and intensity of flowering and seed production which inturn will have negative impact on Forest Genetic Resources. The impact of climate change on mangrove genetic resources are also indicated in the several countrys' forest reports.

vii. Climate impact on species and ecosystem

Any significant changes in temperature and precipitation due to climate change will have significant impact on species composition and forest cover. The fire and pest out break may alter the physiological process there by change the genetic pattern. Due to changing climates, the range of some species may expand and the other may diminish which may create paradigm shift in forest types. There are evidences of migration of keystone ecosystem in Russia. It is also expected that boreal forest may shift towards the pole.

viii. Epidemic status of pest and diseases

The climate change issues and the associated changes in temperature will have a serious influence on the incidence and spread of both pest and disease. In certain cases, the pest and disease out breaks are very serious and may lead to epidemic status thus resulting in complete failure of plantations.

S.No	Major Pest and Diseases	Causal Organisms	Impact
1.	Mountain pine beetle	*Dendroctonus ponderosae*	14 million ha of boreal forest affectin North America
2.	Fungus	*Heterobasidion parviporum*	Root and Butt rot in coniferous forest in finland
3.	Gall Wasp	*Leptocybe invasa*	Complete damage of Eucalyptus in India

ix. Change the fire regimes

The climate change is believed to alter the fire occurrence in frequency and intensity. The warm temperature and drought conditions in Alaska created wild fire in over 2.5 million ha. The increasing fire rate and intensity is almost equivalent to cleaning the total forest and seriously affect the Forest Genetic Resources. Adverse fire may directly affect the biological dispersal agent and may cause the migration of genes between populations.

x. Invasive species

One of the major threats indicated for the loss of Forest Genetic Resources is the role played by spread of invasive species including plants, insect pests and microbial pathogens. The invasive plants due to their rapid seed production followed by colonization will invade natural and slightly distributed forest association and gradually become the dominant species of the locality replacing the natural vegetation. This may lead to loss of several species. *Prosopis juliflora* in Africa and India is the major invasive species and witnessed considerable negative impact on native tree population both in terms of species and genetic diversity. The spread of *Lantana camara* in Western Ghats is also reported to alter the native vegetation. The tropical American tree *Miconia calvescens* has become one of the world's most invasive species and has spread more than quarter of a moist tropical forests of France.

xi. Genetic pollution

Genetic pollution is also identified as a major drivers of change in Forest Genetic Resources but its effects are mostly undocumented. However it is forecasted that mixing of gene pool may results in inter breeding with local population with nonlocal germplasm which may lead to loss of desirable characters and lead to reduced local adaptation in subsequent tree generation. Similarly mixing of gene pools may lead to incorporations of undesirable genes resulting in diminishing economic value of production forest.

CONSERVATION OF FOREST GENETIC RESOURCES

The conservation of Forest Genetic Resources has received potential attraction both nationally and globally due to their productive, economic, environmental and other service roles. The conservation of FGR is approached by two methods viz., *In-situ* and *Ex-situ* conservation methods which are discussed below.

I. *In-situ* conservation

It means conservation in the natural, original place or position. It includes conservation of plants and animals in their native ecosystem where they naturally occur. This is commonly practiced for wild flora and fauna and not to the domesticated plants and animals. This type of conservation is aimed at protection and preservation of land races, wild relatives, wild forms etc.

Eg: **National Parks, Sanctuaries, National monument, Cultural landscape, Biosphere reserve etc.**

a. **National Park:** It is an area dedicated to conserve the scenery, natural objects and the wildlife therein. In National parks, all private rights are nonexistent and all forestry operations and other alternatives such as grazing of domesticated animals are prohibited. However the general public can enter into the National Parks for observation and study purpose. The tourism is also allowed in certain parts provided that it is not disturbing the animals. In India, 103 National Parks are established which serve the purpose of *in-situ* conservation.

b. **Wildlife Sanctuary:** It is an area dedicated to protect the wildlife but it considers the conservation of species only and also the boundary of it is not limited by state legislation. In sanctuary, hunting or capturing of any species of birds and mammals is prohibited except by or under the control of the highest authority in the forest department responsible for the management of sanctuary. Private ownership, forestry and other practices are allowed to the extent that these activities do not adversely affect the wildlife. In India, there are 531 wild life sanctuaries which act as a major *in-situ* conservation areas.

c. **Biosphere Reserve:** It is an area created to conserve the biological diversity and genetic integrity of plants, animals and microorganisms in their totality as part of the natural ecosystem to ensure perpetuation and unhindered evolution of the living organism.

II. *Ex-Situ* Conservation:

Ex-situ conservation is the preservation of components of biological diversity outside their natural habitats. This involves conservation of genetic resources, as well as wild and cultivated or species, and draws on a diverse body of techniques and facilities. Such strategies include establishment of botanical gardens, zoos, conservation strands and gene, pollen, seed, seedling, tissue culture and DNA banks.

i. Seed gene bank:

These are cold storages where seeds are kept under controlled temperature and humidity for storage and this is easiest way to store the germplasms of plants at low temperature. Seeds preserved under controlled conditions (minus temperature) remain viable for long durations of time.

ii. Gene bank:

Genetic variability also is preserved by gene bank under normal growing conditions. These are cold storages where germplasm are kept under controlled temperature and humidity for storage; this is an important way of preserving the genetic resources.

iii. Cryopreservation:

This is the newest application of technology for preservation of biotic parts. This type of conservation is done at very low temperature (-196°C) in liquid nitrogen. The metabolic activities of the organisms are suspended under low temperature, which are later used for research purposes.

iv. Tissue culture bank:

Cryopreservation of disease free meristems is very helpful. Long term culture of excised roots and shoots are maintained. The meristem culture is very popular in plant propagation as it's a virus and disease free method of multiplication.

v. Long term captive breeding:

The method involves capture, maintenance and captive breeding on long term basis of individuals of the endangered species which have lost their habitat permanently or certain highly unfavorable conditions are present in their habitat.

vi. Botanical gardens:

A botanical garden is a place of ex-situ conservation area which harbours wide range of trees, shrubs, herb and grass. The plants of flowers, fruits and vegetables are also grown. The botanical gardens provide beauty and calm environment. Most of them have started keeping exotic plants for educational and research purposes.

vii. Animal Translocation:

Animal translocation is also one of the *ex-situ* conservation methods which is in practice for most endangered and endemic animals.

Translocation is carried in following cases:

i. When a species on which an animal is dependent becomes rare.

ii. When a species is endemic or restricted to a particular area.

iii. Due to habit destruction and unfavorable environment conditions.

iv. Increase in population in an area.

viii. Zoological Gardens:

In zoos, wild animals are maintained in captivity and conservation of wild animals (rare, endangered species) is carried out. The oldest zoo, the Schonbrumm zoo which exists today also, was established in Vienna in 1759.

In India, the 1st zoo came into existence at Barrackpore in 1800. In the world, there are about 800 zoos. Such zoos have about 3000 species of vertebrates. Some zoos have undertaken captive breeding programmes also.

Advantages of *Ex-situ* preservation:

i. It is useful for conserving the declining population of species.

ii. Endangered animals on the verge of extinction are successfully bred.

iii. Threatened species are bred in captivity and then released in the natural habitats.

iv. *Ex-situ* centers offer the possibilities of observing wild animals, which is otherwise not possible.

v. It is extremely useful for conducting research and scientific work on different species.

SUMMARY

The Forest Genetic Resources are valued for their economic, environmental and ecosystem benefits. However these FGRs are under tremendous pressure due to the rapid rate of deforestation, habitat alteration and indiscriminate extraction of forest products. These activities have endangered large number of species and some of them have drive to a stage of extinction. The economically important tree species are also exploited continuously which lead to the presence of only emaciated population. All these threat demands sustainable conservation and management of Forest Genetic Resources. Hence this chapter presented the scope and potential of Forest Genetic Resources along with recent developments in commercially important genetic resources improvement and conservation programmes. This chapter also identified the potential causes for the threat of genetic resources and conservation methods to protect these genetic resources both within the ecosystem and outside its natural ranges.

12

CLONAL FORESTRY

INTRODUCTION

Clone based plantation development programme has been emerged as an alternate to traditional seed based zygotic breeding programme due to its role in productivity improvement along with reduction in rotation. Trees in general are highly heterozygotes and heterogenous and exhibit greater variation in the field which results in uncertain wood and biomass production. Several plantation program established even with improved seed sources have exhibited lower production of less than 10 m^3 / ha / annum and require long growing period. This is attributable to highly variable individuals even within the improved seed sources. This is true because of the cross pollination nature of tree species and the associated heterozygosity in the individuals of the population.

Under such circumstances, clonal forestry has gained significant attention and attraction from 1990 onwards and exhibited greater success in species like eucalyptus, poplars, salix, pinus and even timber trees like teak, gmelina and shisam. The productivity levels of over 25 m^3 / ha / annum has been achieved in eucalyptus through clonal propagation from across wide range of soil and climate regions. This productivity improvement coupled with genetic uniformity and short duration attracted wood based industries, forest department, research organizations and tree growing farmers towards development and deployment of clonal technology in a wide range of tree species.

Currently over fifteen different tree species like Casuarina, Eucalyptus, Poplars, Melia, Shisam, Teak, Kadam, African mahogany, Mahogany, Toon, Jatropha, *Morus alba,* Bamboos,

Subabul, Acacia hybrid etc., are exploited through clonal technology and play a significant role in meeting the raw material requirement for both domestic and industrial wood requirement. This chapter elaborates the major technological innovation and interventions in clone based plantation development programme in major and commercially important tree species.

DEFINITION OF TERMINOLOGIES

Clone :
- The aggregate of individual organisms, originating from one sexually produced individual and maintained exclusively by asexual propagation.
- A group of vegetatively-propagated organisms consisting of an ortet and its ramets.
- A cell line of single-cell origin.

Clonal Forestry :
- Clonal forestry refers to the large-scale deployment of relatively few (10 to 50), known-superior clones that have proven their superiority in clonal tests.
- Clonal forestry can be defined as 'the deployment of tested clones'.

Ortet :
- The initial individual (usually from a zygotic embryo) that is vegetatively propagated to produce a clone.
- The original single parent plant from which a clone ultimately derives.

Ramet :
- A vegetatively reproduced copy of a plant.
- All vegetative propagules of an ortet are ramets.
- Any of the individuals in a group of clones is referred to as a ramets.

True to type :
- Individuals having essentially the same phenotype as the donor plant, cultivar or clone.

Grafting :
- Grafting is the process of joining two plants together (an upper portion and a lower portion) to grow as one. The upper portion of the plant is known as the scion, which is

attached to the lower portion known as the rootstock.

- Grafting or graftage is a propagation technique whereby tissues of plants are joined so as to continue their growth together. The upper part of the combined plant is called the scion while the lower part is called the rootstock.

Layering : Layering is a method of propagating a plant in which a shoot is fastened down to form roots while still attached to the parent plant.

Budding : Budding is a type of asexual reproduction in which a new organism develops from an outgrowth or bud due to cell division at one particular site.

CLONAL PROPAGATION METHODS

Generally plants and trees are multiplied vegetatively through the following methods

i. Rooting of cuttings
ii. Grafting
iii. Layering
iv. Budding
v. Tissue culture

i. ROOTING OF CUTTINGS

This is most predominantly used method and deployed in wide range of propagation method like stem cuttings, root cuttings, leaf cuttings, leaf bud cuttings, softwood cuttings, semi hardwood cuttings and hardwood cuttings etc., The respective cuttings from the identified mother plants are separated out and are treated with rooting hormones like IBA, IAA are incubated under green house or poly house conditions. The rooted cuttings are then used as a clonal plant for all practical applications.

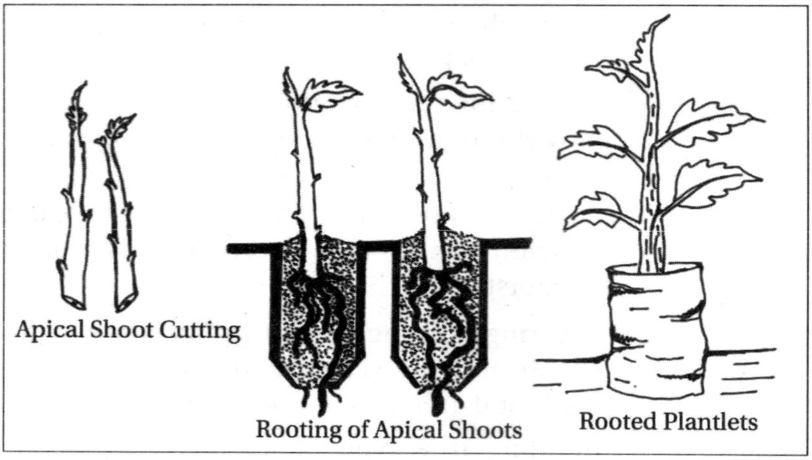

Fig. 12.1. Rooting of cuttings

ii. GRAFTING

It is another important clonal propagation particularly for fruit trees which deals with joining plant parts (Scion and Stock) together in such a way that these plant parts readily unite and continued to grow as one plant. The scion obtained from superior genetic stock is grafted on to the root stock of the seedling plant of the local variety. The upper portion of the grafted plants is called scion and the lower portion of the grafted plants is called stock and the success of the grafting depends on the compatibility of scion and stock along with the intact cambial contact.

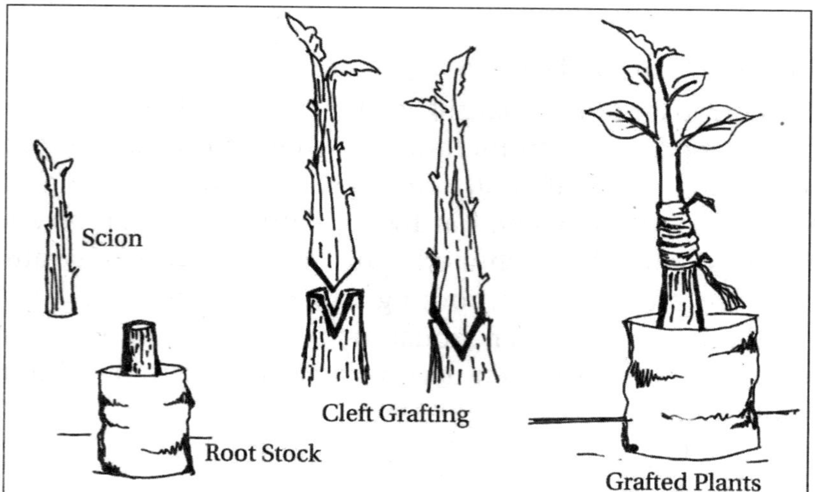

Fig. 12.2. Grafting of tree species

iii. LAYERING

It is a method of auto vegetative propagation wherein roots are induced on the girdled branches while they are still attached to the tree. The identified branches in the tree are girdled by removing the bark in a ring shape followed by placing the potting media around the cut portion with or without treatment of rooting hormones. After placing the potting media, the entire portion is tied with moisture proof wrapper. After rooting in the girdled area, the rooted branches are cut separately grown in container for certain period which are then used for planting programme. Again this method is predominantly used for fruit and flowering trees.

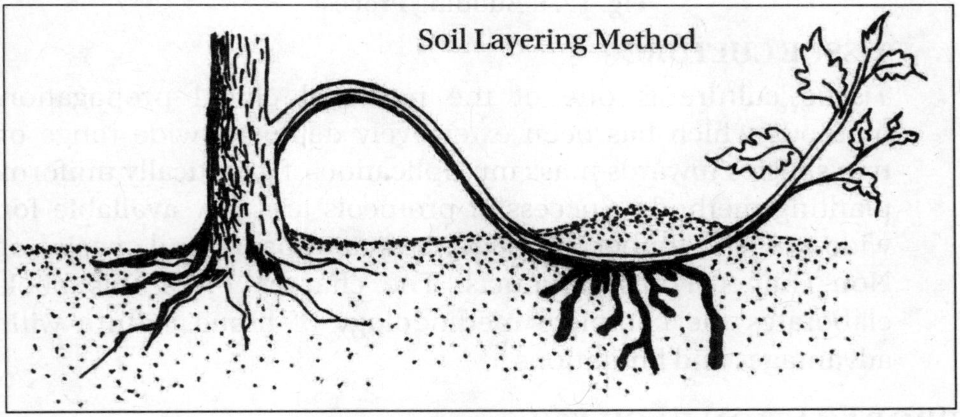

Fig. 12.3. Layering methods

iv. BUDDING

It is one of the heterovegetative propagation wherein the buds from the superior trees are collected and budded on the trees of interest. After successful union, the successfully budded portion is separated out and placed in the containers for further growth and development and deployed in plantation programme.

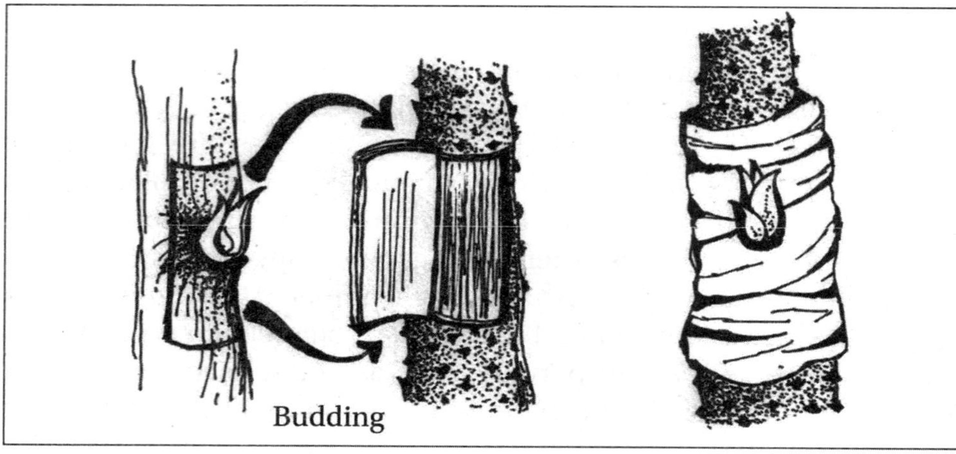

Fig. 12.4. Budding Process

v. TISSUE CULTURE

Tissue culture is one of the potential clonal propagation methods which has been extensively deployed wide range of tree species towards mass multiplication of genetically uniform planting methods. Successful protocols are now available for wide range of timber, plywood, pulp and paper and species of Non-Timber Forest Products. The chapter 14 of this book elaborates the complete methodology of tissue culture with advantages and limitations.

PROPAGATION FACILITIES

The success of clonal technology depends on the establishment and availability of suitable infrastructure facilities coupled with the mother clonal garden. The following are the major essential infrastructure needed for clonal multiplication process.

i. Clonal Mother Garden

ii. Mist Chamber or Green House

iii. Hardening Chamber

iv. Open Nursery

i. Clonal Mother Garden

It is an essential establishment which houses the mother plants in cement trough or FRP sheets. The detailed establishment along with the size and filling materials are presented in the mini clonal technology.

Fig. 12.5. Clonal Mother Garden at FCRI, Mettupalayam

ii. Mist Chamber or Green House

- The clonal propagation of several tree species depends on the availability of controlled growing conditions towards cell division and cell differentiation.
- The temperature, humidity, moisture regime, air circulation and free from infestation are the major environmental requirements for propagation of tree species through clonal technology particularly using rooting of cuttings. This requires an establishment of full fledged mist chamber or green house.
- The green house should be established with cement concrete base and erected by using steel pipes covered by UV stabilized poly film with 200 micron thickness. This poly film should cover the side walls and the roof of the green house.
- The backside of the green house should be assembled with a cool seal pad system which facilitates movement of external air in to the green house with a reduced temperature.
- The front end of the green house should be fitted with one or two exhaust fans depending on the size of the green house in order to drive away the heat air inside the green house and to suck the air passed through the cool seal pad system to maintain the temperature inside the green house.
- The entire working area of the green house should be assembled with misting or fogging system with a timer facility to control the humidity and moisture requirement of the propagated clones.
- The green house should be connected with the continuous supply of quality water with neutral pH.

Fig. 12.6. Green house at FCRI, Mettupalayam – Capacity 0.5 million per annum

iii. Hardening Chamber

- The third major infrastructure requirement for a propagation facility is the establishment of the hardening chamber. It is otherwise called as a shade net house which is established with RCC base, supported by steel frames.
- The entire housing area is to be covered with 50% or 70% shade net in order to protect the rooted plants from direct sunlight.
- The entire working area of the shade net house has to be provide with misting facilities to maintain the humidity and the water requirement of the rooted plants placed under hardening process.

Fig. 12.7. Hardening chamber / Shade House at FCRI, Mettupalayam

iv. Open Nursery

- The clonally propagated plants demand suitable hardening process by exposing them slowly to varied atmospheric conditions

- Hence establishment of open nursery with a provision for automatic sprinkler system with housing facility to place the clonal plants are very important for secondary hardening of clonal plants.
- The clonal plants once acclitamatized in open nursery condition are amenable for planting out in the field towards successful establishment of clonal plantations.

Fig. 12.8. Open Nursery at FCRI, Mettupalayam

MINI CLONAL TECHNOLOGY

Though wide ranges of vegetative and clonal propagation methods are available for application but only the rooting of cuttings is deployed commercially across wide range of species towards mass multiplication of quality planting materials. The pulp and paper industries across the world almost completely rely on this technology to mass multiply the species like eucalyptus, casuarina, subabul, acacias, bamboos, melia etc., This rooting of cuttings based clonal technology has been further refined and the concept of mini clonal technology is replacing the traditional hedge garden based clonal technology which is described in detail in this chapter.

Clonal selection programme

The first stage in Clonal propagation is selection of a clone. It is the selection of the individual trees based on its morphogenic superiority from the existing plantations. The individual trees are selected based on growth attributes, tree morphology, tree health and the apical dominance. In general, trees are selected as early in 3 years after planting and for many species selection in half rotation age is also practiced. The clonal selection also depends on the species, objectives of management and also the end utility coupled with the commercial value.

Once the trees are selected, the trees are cut at base preferably at 15 cm above the ground level and are allowed to coppice. These coppice shoots are harvested and treated with rooting hormone and ramets are multiplied for field testing and evaluation. Simultaneously, the harvested wood is fragmented into wood samples and is analyzed for physical, chemical, mechanical and anatomical properties. The wood properties viz., cellulose content, lignin content, kappa number, specific gravity and fibre length are characterized for screening the superior most wood quality clone. Based on the field productivity coupled with wood quality, the selected clones are finally screened and are deployed for mass multiplication through mini Clonal technology.

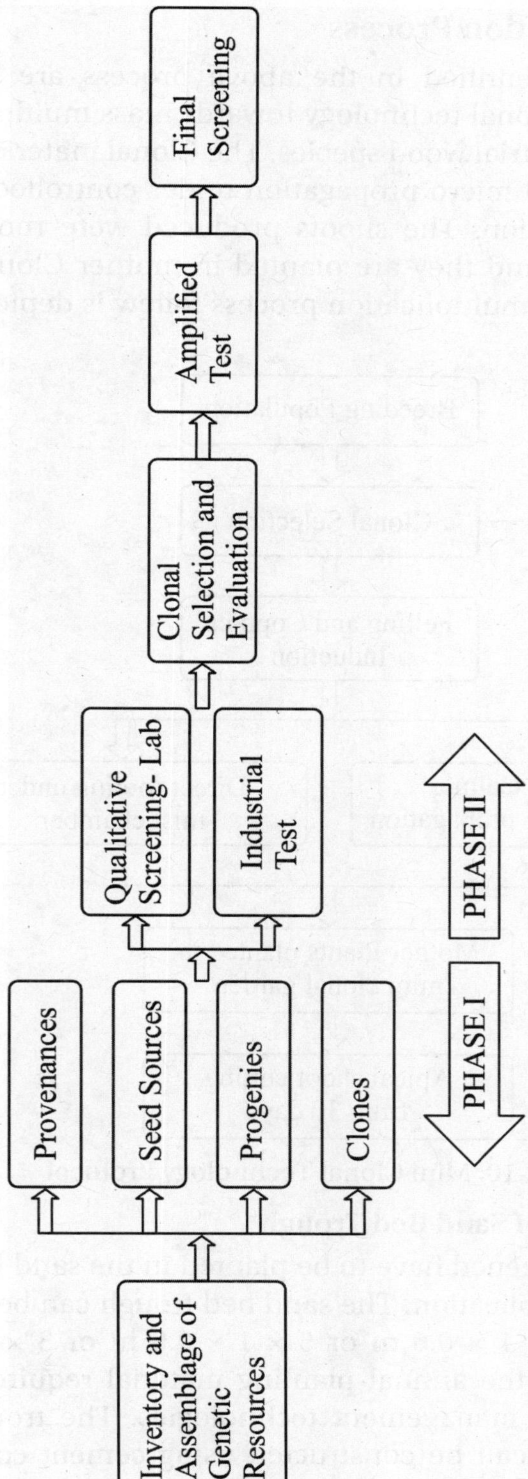

Fig. 12.9. Clonal Selection and Evaluation

Clonal Multiplication Process

The clones identified in the above process are multiplied through the mini clonal technology towards mass multiplication of domestic and industrial wood species. The Clonal materials are first established through micro propagation under controlled light and temperature condition. The shoots produced were rooted under invitro conditions and they are planted in mother Clonal garden. The general Clonal multiplication process follow is depicted in Fig. 12.10.

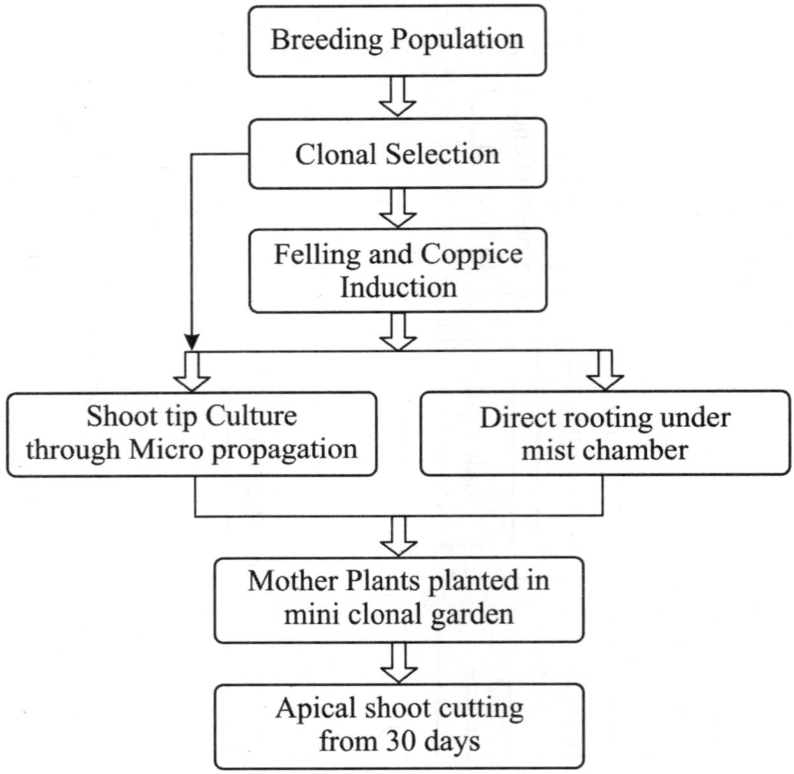

Fig. 12.10. Mini Clonal Technology Protocol

a) Construction of Sand Bed Trough

The clones screened have to be planted in the sand bed trough for mass multiplication. The sand bed trough can be of various sizes viz., 10 × 1 × 0.6 m or 5 × 1 × 0.6 m or 3 × 1 × 0.6 m depending on the annual planting material requirement and the associated management technologies. The trough of the required sizes can be constructed using cement concretes or fabricated through GI sheets. Once the bed size is designed and established, the bottom of the trough is filled with 20 mm stones upto the height 20 to 25 cm from the base upwards. Then the portion above stone filling should be filled with fine sewed river sand. The trough should have facility for proper drainage at both the ends. The trough can be covered with 100 micron UV stabilized poly ethylene film on the top and covered with insect proof mesh along the sides in order to protect the plants from pest and diseases. The design of a trough is furnished in fig 12.11. The bed should be properly connected with drip lines to facilitate irrigation and fertilization.

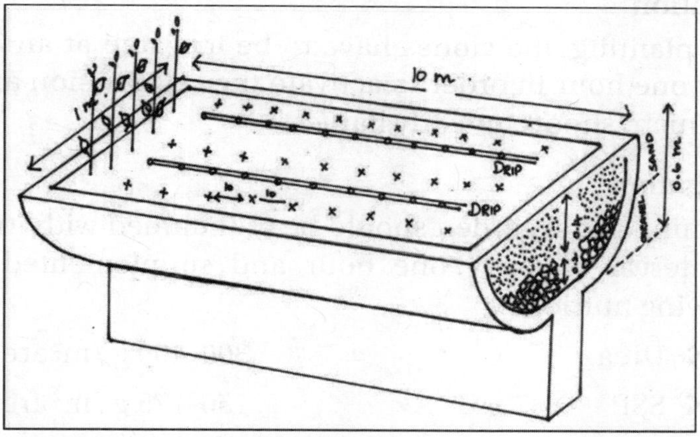

Fig. 12.11. Design of a Sand Bed Trough - 10 m x 1 m x 0.6 m

b) Planting and management of clones

Once the trough is established, the prioritized clones are planted for mass multiplication. The ramets are planted at an espacement of 10 cm x 10 cm (between rows and between plants) so as to cover 100 plants/m². So in 10 m² plot, about 1000 plants can be planted. The design of Clonal planting in the sand bed trough is furnished in Fig 12.12.

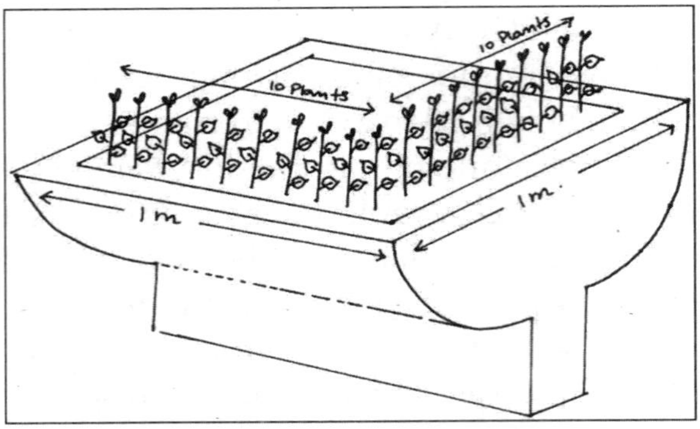

Fig. 12.12. Design of clonal planting in sand bed trough

c) **Irrigation**

After planting, the clones have to be irrigated at an interval of every one hour in order to activate the cell division and induce new micro shoots for early harvest.

d) **Fertigation**

The mini clonal garden should be maintained with irrigation at an interval of every one hour and supplemented with the following nutrients.

- Urea - 300-400 g /m^2 area
- SSP - 150-175 g /m^2 area
- KCL - 175-250 g /m^2 area
- Micro nutrient mixture - 100 g /m^2 area

The nutrients can be applied twice or thrice depending on the rate of growth of plants and the species involved in the multiplication process. Alternatively application of All 17 (NPK) or All 19 (NPK) is also recommended for application.

e) **Clonal Management**

The plants are allowed to grow upto 30-60 days by applying the required nutrient composition. After 60 days, the plants are pruned at required size preferably at half of the plant to induce new shoots.

f) **Harvesting of Apical Shoot Cuttings**

With continuous irrigation and nutrient management, the cut end stem will start producing new shoots from 8-10 days onwards and after 15-20 days the apical shoot cuttings can be collected and treated with 2% carbendazim solution. After each and every harvest, the irrigation and fertilizer should be continued daily to induce next cycle shoots.

g) Mini cutting Treatment

The apical shoot tip cuttings do not require basal hormone treatment for root initiation due to active cell divisions and the presence of balanced hormonal regulation. These shoot tip cuttings form new apical meristems and the leading growing tip does not originate from the lateral buds. Similarly the roots tend to grow straight down from the base of the cutting as opposed to growing from the side of the stem. This resulted in rapid growth and development coupled with vigorous and quality planting material.

h) Planting of Apical Shoot Cuttings

The newly induced shoots were separated from the plants and are planted in 90 cc root trainers filled with decomposed coir pith without any external hormonal treatment. The rooting started in 15 days and 25 days old rooted plants are ready for hardening. Some species require hormonal treatment for which IBA at 1500 – 3000 ppm powder formulation can be applied before planting in the root trainer.

i) Green House Conditions

The root trainers are kept under green house conditions with a temperature regime of 32–35°C and a relative humidity of 85-95%. Periodical watering once in every 30 minutes will ensure early, easy and uniform rooting process.

j) No. of cycles of shoot harvest

Through this intensive mini garden management, at least one harvest per month is possible. For every harvest, a minimum of 5 shoots per plant is ensured for most species barring few exceptions.

k) Acclimatization and Hardening

The rooted plants are hardened in a shade house condition with 50% shading for 7-15 days and maintained with adequate irrigations. After hardening chamber, the plants are lifted to open nursery for 30 days. Watering is done 2 times a day and the fertilizer of all 19 (N:P:K) can be applied at the rate of 5g/plant. During this hardening process, application of carbendazim (2g/l) or triazophos (2ml/lr) is recommended based on the incidence of diseases and pests.

Advantages of Mini Clonal Technology

- Rooting ability of mini clonal technology is High
- Propagules expressed variability due to epigenetic or somatic variations in the existing clonal technology is avoided through the mini clonal technology.
- This technology can produce more number of cuttings per year and per unit area
- It can be maintained upto 5 years
- It does not require external rooting hormone treatment.
- This technology has shown great potential for offering technical and economic advantages.
- Increased uniformity and greatly reduced topophysis effects
- Compared to existing clonal technology, the mini clonal technology has improved rooting potential, rooting speed, quality root systems as well as reduced cost
- Cuttings can be taken from 60 days onwards and after that once in every 15-20 days against the periodic collection in the existing technology

Limitations

- Initial high cost of establishment
- Intensive care and management
- Requirement of skilled labour
- Continued research and development support

MINI CLONAL TECHNOLOGY VS TRADITIONAL CLONAL TECHNOLOGY

The regular Hedge Garden Technology have used nodal cuttings and in many cases shoots and roots tended to initiate from the lateral buds along the stem. This technology demanded large volume of rooting hormone and also time required for production of cuttings is comparatively high. This besides, the large nursery area requirement coupled with the time required for cutting collection process and the existence of topophysis effect are the major bottle necks in the regular Hedge Garden Technology. The current mini clonal technology has exhibited significant advantages over the existing technology in all the issues addressed above. Some of the significant advancements are furnished below.

a) No. of cuttings

The number of cuttings available per annum from 100 m² area through the mini clonal technology is about 5,00,000 plants compared to only 50000 shoots from the existing technology. The comparison is furnished below.

Apical shoot production from Mini clonal technology 100 m² area

No. of plants	No. of shoots harvested /per plant/month/	No. of harvest / Annum	Total No. of shoots available / year
10,000	5	10	5,00,000

Nodal cutting production from traditional hedge garden technology in 100 m² area

No. of plants in 4000 sq mt	No. of shoots harvested /per plant/month/	No. of harvest / Annum	Total No. of stem cuttings harvested per year
100	50	10	50,000

b) Efficiency

The efficiency of mini apical shoot cuttings is significantly higher than the nodal shoot cuttings and the significant differences are furnished below.

SHOOT CUTTINGS	MINI CUTTINGS
• Reduced rooting % • Poor quality root system • Variation in field growth • More gestation period for propagation • Larger areas are required • Root promoting hormones required	• Increased rooting% • High quality root system • Uniform growth in the field • Less gestation period almost half • Less area is required • No growth hormones are required

c) Productivity

The important differences noted in term of productivity between intensively managed mini clonal plant and field-grown stool plant can therefore be explained by the physiological state difference and particularly by juvenility and the active hormonal regulation. Results also indicated that the plants multiplied through mini clone based technology have produced higher root biomass compared to the root biomass of Hedge Garden Technology. This increased root biomass has ensured maximum survival, higher vigour, increased uniformity and higher productivity.

An Ideal Clonal plant

A ideal plant should look like healthy and shows no evidence of physical damage. The plant should have a straight stem with a single leading shoot (or "leader") at the very top which should end in a healthy terminal bud. This will provide the basis of the next year's growth. The plant should be sturdy - that is the width of the stem (the 'root collar diameter') just above the roots should be much stronger than the rest of the stem. The plant should also be well balanced in that there should be sufficient roots to support the upper part of the plant. The root system should contain a large element of fibrous roots. As a rule of thumb, the root system should constitute at least a quarter of the plant parts.

APPROACHES FOR COMMERCIALIZATION OF THE TECHNOLOGY

The technology can be commercialized very easily through systematic approach. It can be commercialized through the following stake holders.

- Through wood based industries across the country
- Nursery and rural industries
- NGOs and self help groups
- Government department like state forest department/ rural development department etc.,
- Members of Consortium of Industrial Agroforestry
- Agroforestry Business Incubators
- Student Entrepreneurs
- Startup Firms

OPERTIONAL CLONAL FORESTRY

The success of vegetative propagation was traced 1500 years back in China which has now commonly pronounced as clonal forestry. Successful development and implementation of clonal forestry is very well witnessed in several hardwood species particularly in the success of Eucalyptus by the Brazilian wood based industry is very significant. The clonal technology based plantation forestry establishment has benefitted through yield realization of 10 to 22% in conifers and about 25 to 50% in hardwood species. Several hardwood species suitable for timber, pulpwood, plywood, dendro power, bio fuel etc., are also improved through clonal technology and successful development of clonal forestry programme in those species are on the rise. All these clonal forestry development programme have adopted several strategies but the most common operational strategies involved in establishment of clonal forestry plants are diagrammatically presented below in Fig. 12.13

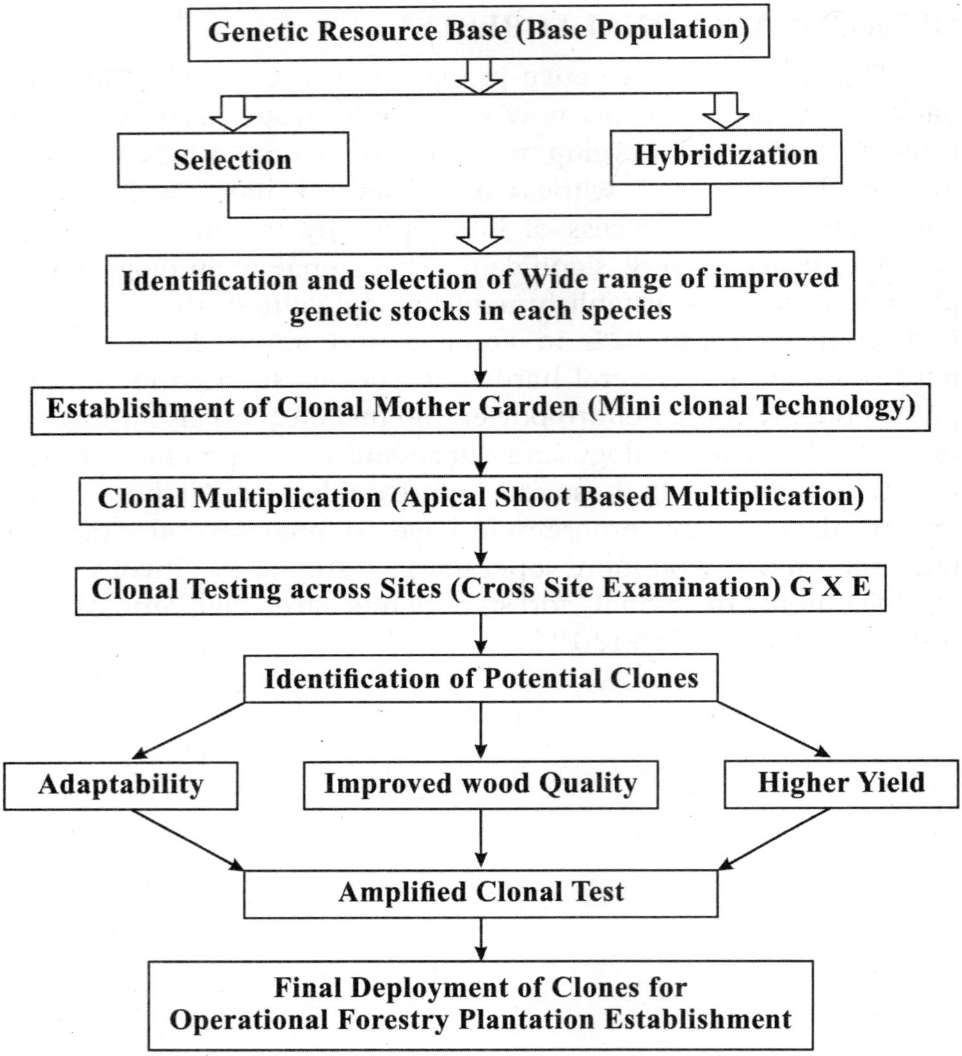

Fig. 12.13. Diagrammatic Representation of Operational Clonal Forestry

ADVANTAGES OF CLONAL TECHNOLOGY

- The clonal technology has an advantage of exploiting natural variations by ensuring complete utilization of additive and non-additive genetic variability.
- This technology helps to mass multiply larger number through less space, rapid multiplication and with minimal cost.
- Most tree crops are cross pollinated and hence, they are highly heterogenous and heterozygous. Under such circumstances,

the Clonal technology has an advantage of exploiting heterozygosity which leads to increased productivity.

- This clonal technology particularly, grafting technology helps to produce clones for specific purposes like early flower and fruit production, broader canopy and short form which address the objective of tree breeders.
- Several tree species, like teak, requires long period to induce flower and fruits which affect the artificial regeneration process for those species. Clonal technology is an excellent alternate method to mass multiply the species and facilitate easy artificial regeneration.
- Clonal technology will lead to fast and immediate genetic gain compared to the long life cycle and low genetic gain from the traditional seed orchard based breeding programme.
- Some of the tree species are naturally sterile and regeneration of these species depends fully on the clonal propagation methods.
- The seeds of many tree species are highly recalcitrant and loss their viability at a short period and is one of the major detracting factors for plantation development programme.
- The clonal technology helps to mass multiply recalcitrant tree species throughout the year and facilitates plantation development programme during any period of the year.
- The grafting based clonal multiplication process helps the breeder through early flower induction and facilitates hybrid development programme along with the associated hybrid vigour or heterosis development.
- The mini clonal technology is a recent advancement which eliminated all the obstacles in the traditional clonal propagation methods.
- This technology ensure rapid multiplication, ensure genetic uniformity, eliminate epigenetic variation and increased the vigour and productivity.

LIMITATIONS OF CLONAL TECHNOLOGY

- Clonal technology involves huge initial investment for the establishment of propagation and hardening chambers.
- This technology needs skilled labourers to handle the activities of clonal selection, clonal production, clonal hardening and maintenance of clonal mother garden which demand specific training and knowledge to be inculcated to the propagator on a sustainable basis.
- The clonal plants do lack tap root system which may lead to poor anchorage and are susceptible to uprooting during the period of heavy winds.
- Another major problem is, clonal technology differs from species to species and even within species it varies from clone to clone which necessitates continuous development of clone specific or species specific multiplication technology.
- The clonal technology and the associated clonal forestry demand continuous entry of new clone to create diversity; otherwise, the mini clonal plantations will have more susceptibility towards biotic and abiotic factors.

SUMMARY

Traditional seed orchard based breeding program has paved a significant contribution towards productivity improvement and generation of quality timber. However, traditional breeding program has been delimited due to long rotation of several tree species coupled with low genetic gain realized through the generations of tree breeding. Under such circumstances, the clonal technology is a break through intervention which exploits the existing variations and results in rapid genetic gain incorporating the advantage of the inherent genetic qualities of the clones. This genetic gain with desirable quality and also with genetically uniform identicals leads to increased productivity improvement. The complete utilization of additive and non- additive genetic variation is yet another advantage of clonal technology. Systematic and decentralised cross site examination of diverse clones will help to identify site specific adaptive clones to cater to the needs of various wood based industries. It is also possible; this technology will also extend the scope of identifying superior clones for wood quality and yield, resistance clones for biotic and abiotic disturbance and also utility based clones. Hence, the clonal technology and associated clonal plantation development are more significant towards meeting the raw material requirement of both domestic and industrial demand besides acting as an excellent source of income and employment generation activities to the people. The recent advances made in clonal technology, particularly the mini clonal technology established for Eucalyptus, Casuarina, Melia, African mahogany and mulberry extend a greater scope of adoption of this technology for wide range of species which may create a paradigm shift in industrial forestry plantation and agroforestry programme of the country towards realising higher productivity and augment the socio economic status of industrial agroforestry practising farmers.

13

EXOTIC FORESTRY

INTRODUCTION

The growing demand for wood and wood products coupled with changes in legal and policy guidelines have restricted the supply of wood and wood products from the natural reserves and protected forest. The industrialization, urbanization, population growth and the changes in housing policy have increased the demand for wood and wood products at an alarming rate. Above all, the increasing use of paper and paper boards in all sectors of industries has accelerated the demand for wood and wood products. This has ushered in a total mismatch between demand and supply of wood and wood products. All these developments have necessiciated large and reliable supply of inexpensive forest products with good wood quality. The existing natural forests due to their poor productivity and legal restrictions on felling operation, could not meet the current escalating demand of wood and wood products. One of the most significant ways to improve the supply of wood and wood products is to deploy and extend the use of the exotic species which are fast grower, easily amenable for varied edapo climatic conditions and supply huge volume of biomass in shortest possible time. Wide range of exotic species like Eucalyptus, Casuarina, Acacias, Silver oak, Poplars, Pines etc., have been introduced into India and are catering the raw material needs of both domestic and industrial sector. However these exotics have exhibited wide range of positive as well as negative deliberations which in most cases detract their promotion. This chapter elaborates the use of exotics species along with their advantages and disadvantages which may be of great use to the readers of the book and practising foresters and wood based industries.

DEFINITIONS

- ❖ An Exotic is a plant that is not native.
- ❖ Exotic is strictly defined as "An introduction from a foreign country".
- ❖ Exotic is one which is grown outside the limits of its natural range.

REASONS FOR WIDESPREAD USE OF EXOTICS

- ❖ In tropics and subtropics, the exotic coniferous timbers have a great need to their superior wood quality because the indigenous forests frequently do not contain the required quality wood.
- ❖ The productivity of native species is very low of 0.5 m^3/ ha/ annum but the exotic plantations have the production potential of 25-30 m^3/ ha/ annum and hence they are preferred predominantly for productive industrial wood plantations.
- ❖ Exotics have wider adaptability and multipurpose utility compared to the indigenous species which are attracted by several forest based industries.
- ❖ Most of the exotics are fast growers which resulting in lowering of the rotation period and increased economic returns. Some of the exotics like Casuarina and Eucalyptus are harvested in less than 5 years wherein no indigenous species are harvested within this period.
- ❖ The tropical and indigenous hard wood forests are so variable and are difficult to manage silviculturally than the exotics. On the other hand, exotics are easy to manage both silviculturally and ecologically due to their short rotation, fast growing and multiple utility. Hence exotics are preferred in place of native species
- ❖ Most foresters generally prefer to use species that are more uniform and whose products are known and accepted. Hence the exotics are widely accepted as a species with uniformity and easy to manage.
- ❖ Marketing potential and manufacturing technology favours the use of exotic woods such as Pines, Poplars, Eucalyptus etc., compared to indigenous species.

- ❖ Availability of quality seed and planting stock is a key to success in any plantation forestry programme. The seeds of the indigenous species are sometimes hard to obtain due to poor understanding of reproductive biology. On the other hand, several exotic species have been developed into improved genetic stocks resulting in better quality and more uniform trees from which seeds are easily available.

CHARACTERISTICS OF EXOTICS

i. An exotic should produce useful timber or other valuable forest products.

Eg: Eucalyptus, Pines, Silver oak, Poplar

ii. The exotic should have ability to colonize the barren ground without shelter. Any land which is barren due to fire, volcanic eruption, bad agricultural practices, exposure to soil and wind erosion are easily colonized by most successful exotics.

Eg: *Prosopis juliflora, Acacia tortilis*

iii. The exotic should be a fast grower and capable of rapid early growth as an individual. However, it should be of high volume producer when grown as a plantation.

Eg: Casuarina, Eucalyptus, Subabul

iv. The exotic should have a capability to express good growth habit taking the advantage of favourable growing period.

Eg: Eucalyptus has an outstanding growth habit utilizing the advantage of favorable growing conditions.

v. Exotics should have multitudinal branching habit and more number of shoots. These branches and shoots produce shoots continuously as long as the growing condition persists. This will help to create a large and structurally sound relief system very quickly and results in increased biomass production.

Eg: Eucalyptus, Poplar, Pines

vi. Exotics should have a strong apical dominance which enables the trunk and the branches to extend their normal axis of development irrespective of damages caused to them. In case, if the leading shoot is damaged, the capability of apical dominance of exotic species permits the continuous growth of vertical axis from a favorably placed leaf axle close to the point where the leading shoot has died. This capability of apical dominance ensures production of clean and straight bole in most exotic species.

Eg: Eucalyptus, Poplar

vii. The exotics should have vigorous root system which will have protection against all harmful agents.

Eg: Prosopis

viii. The exotics should have adequate and appreciable variation which will help to identify variety suitable for different environments.

Eg: Eucalyptus, *Acacia auriculiformis*, Silver oak

ix. The exotics should have capacity to produce adequate seeds or should be amenable for vegetative multiplication which will ensure large scale commercial production.

Eg: Eucalyptus, Casuarina, Acacia mangium

ENVIRONMENTAL CONSIDERATIONS FOR INTRODUCTION OF EXOTICS

It is always better to introduce an exotic to a locality which has a climate and soil similar to that of its home range.

i. Water and rainfall

Water is one of the most significant factors to be colonized before introduction of exotics. For this purpose, rainfall as a climatic index should be carefully monitored for introduction of exotics. Normally, the distribution of rainfall between the cooler and warmer season should be taken in to account than the total rainfall. It is also very important to consider that the total rainfall is adequate for the given species in that particular locality.

If an exotic species is removed from summer rainfall locality to winter rainfall locality, the success of species is more compared to moving a species from winter rainfall zone to summer rainfall zone particularly in the lower altitudes. However, frost is a critical factor in the introduced cooler region which needs to be considered before introducing exotics.

ii. Humidity

Humidity particularly high and low humidity play a critical role in success and failure of exotic species. The species moved from locality experiencing low atmospheric humidity tend to suffer a lot, if it is introduced to a region with high humidity. However, when the species moved from high atmospheric humidity to low atmospheric humidity, there is no much variations in survival and growth.

iii. **Heat and temperature**

Heat is yet another critical factor to be considered before introduction of exotics. If a species is moved to a region where in the extremeness of heat are significantly greater than their natural habitat, then the species growth is likely to be deformed.

iv. **Wind**

Similarly, exotic introduction is also a problem, if the area witnessed heavy and dry wind, than their natural habitat.

v. **Latitude**

Most exotics have a wide latitudinal range of natural occurrences. These exotics when introduced to other countries, better results are obtained by selecting equivalent climates from a higher altitude at a lower altitude than from a lower altitude at a higher altitude. The species selected from higher altitude when introduced to a lower altitude will have longer growing season and produce better mean annual increment. However, the trees of lower altitudes are least influenced by day length.

vi. **Soil**

Soil is the most critical factor for achieving favourable growth and development. Though most exotics, exhibited satisfactory growth and development in good soil coupled with adequate irrigation but the potential of exotics is not confined to only good soils. The artificial supplementation of inorganic nutrients has resulted in very good growth and development of exotics.

Major Exotics introduced into India

Over years, India witnessed introduction of wide range of exotics for varied purpose like fuel wood, ornamental, flowering, shade tree and industrial wood. The major exotic species introduced into Indian sub continent are furnished in the table 13.1.

Table 13.1 MAJOR EXOTIC SPECIES INTRODUCED INTO INDIA

S. No	Common name	Scientific name	Origin	Year of Introduction	Purpose	Source
1.	Mysore gum	*E. tereticornis*	Australia	1790	Pulpwood	Shyam Sundar, 1984 & Wilson, 1972
2.	Rose gum	*E. grandis*			Firewood, small timber, poles	
3.	Grey iron bark	*E.paniculata*				
4.	Spotted gum	*E.maculata*				
5.	Lemon scented gum	*E. citriodora*		1969		
6.	Blue gum	*E. globulus*		1843	Pulp wood and fuel wood	
7.	River Red gum	*E. camaldulensis*		1968	Pulpwood	
8.	Poplar/ Eastern cotton wood	*Populus deltoides*	North America	1950's	Multipurpose tree	Naithani & Nautiyal, 2012
9.	Beef wood	*Casuarina equisetifolia*	Australia and coastal south east Asia	1850's	Firewood	Ray, 1971
10.	Beef wood	*Casuarina junghuhniana*	Indonesia	1950's	Pulp and paper	Ray, 1971
11.	Subabul	*Leucaena leucocephala*	Hawaii, USA	1970'S	Multipurpose tree/ fuelwood	Lohani, 1979
12.	Australian wattle	*Acacia auriculiformis*	North of Australia	1950's	Plantation	Pinyopusarerk, 1990

S. No	Common name	Scientific name	Origin	Year of Introduction	Purpose	Source
13.	Israeli babool	*Acacia tortilis*	Israel	1958	Afforestation programme in arid region	Muthana and Aroara, 1973
14.	Mesquite	*Prosopis juliflora*	Mexico, South America and Carribean	1857	Afforestation	Gomes, 1961 & Duthie, 1960
15.	Black wattle	*Acacia dealbata*	Australia	1840	Firewood	Lorenzo *et al.*, 2010 and Bhat *et al.*, 2012
16.	Silver oak	*Grevillea robusta*	Australia	19th century	Shade tree	Luna, 2005
17.	Radiata Pine	*Pinus radiata*	Central Coast of California and Mexico		Pulp and lumber	
18.	Brown salwood	*Acacia mangium*	North eastern Australia, Indonesia and Papua guinea	Mid 1960's	Reforestation purpose	National Research Council 1983, & Pinyopusarerk *et al.* 1993
19.	Physic nut	*Jatropha curcas*	Central America and Mexico	16th century	Multipurpose shrub	Heller, 1996

S. No	Common name	Scientific name	Origin	Year of Introduction	Purpose	Source
20.	Gulmohar	*Delonix regia*	Madagascar.	1840	Ornamental tree	Anon,1983
21.	Paradise tree	*Simarouba glauca*	Southern florida, West Indies and brazil	1966	Plantation	Armour, 1959 , Juyal *et al.* , 1991 & Rath, 1987
22.	Annatto	*Bixa orellana*	America		Medicinal pursposes	Bole,1995
23.	Black wattle	*Acacia mearnsii*	Australia	Mid 19th century	plantation	Luna, 2005
24.	Indonesian Albizzia / Moluccan Albizzia	*Albizia falcataria*	Indonesia	Mid 1970's	Shade tree	Luna, 2005
25.	Paper mulberry	*Broussonetia papyrifera*	China and Japan		Paper printing	Guha *et al.*, 1964
26.	Japanese cedar	*Cryptomeria japonica*	Japan	1862	Mixture plantation	Luna, 2005
27.	Balsa wood	*Ochrama pyramidalis*	America	1800	Plantation	Anon, 1981
28.	Rubber tree	*Hevea brasiliensis*	Brazil	1873	Rubber	Hilton, 1995
29.	Darwin silky oak	*Grevillea pteridifolia*	Australia		Reclaim bauxitemined areas	Boland and Tumbull, 1989 & Prashad, 1988

S. No	Common name	Scientific name	Origin	Year of Introduction	Purpose	Source
30.	Jerusalem thorn	*Parkinsonia aculeate*	Mexico		Afforestation of dry areas	Anon, 1983
31.	Caribean pine	*Pinus caribaea*	Br. Honduras	1970's	Plantation	Ghosh, 1877
32.	Patula pine	*Pinus patula*	Mexico	1975	Plantation	Ghosh, 1977
33.	False acacia	*Robinia pseudoacacia*	North America	1890	Reforest barren hills	Muthoo and Kango, 1965
34.	Rain tree	*Samanea saman*	Central America	1880	Road side avenue	Anon, 1983
35.	White willow	*Salix alba*	Europe	1916-17	Cricket bats	Luna, 2005
36.	Broad leaved sweiteinia	*Swietenia macrophylla*	Central America		Ornamental	Anon, 1981
37.	Small leaved Mahogany	*Swietenia mahogani*	Cuba, Jamaica	1995	Ornamental avenue tree	Luna , 2005
38.	Jacaranda	*Jacaranda mimosaefolia*	Brazil and Argentina	1841	Ornamental tree	Luna, 2005
39.	Indian Lavender	*Bursera peniciata*	Mexico	1960	Plantation	Luna, 2005
40.	Cashew nut	*Anacardium occidentale*	Tropical America	Between 1563 and 1570	Nut	Luna, 2005
41.	Australian black wood	*Acacia melanoxylon*	Tasmania, south Australia and Queensland	1840	Avenue tree	Luna, 2005

SUCCESSFUL EXOTIC SPECIES IN INDIA

During the last ten decades, wide range exotic species have been introduced in to India for various purposes. Some of the exotic species are introduced for ornamental purpose but has become commercial importance due to the fast growing nature coupled with the amenable wood quality. Among several exotic species introduced, few species like poplars, eucalyptus, casuarina, silver oak, wattles and tropical pines have gained significant importance due to their multifarious utility. These species are now domesticated and cultivated on a larger area across the country which accounts to be over 5 million hectare. These species are contributing almost over 70% of the wood requirement of the country. The salient features of most successful exotic species in India are furnished below.

1. Poplars

Poplars played a significant role in meeting the raw material demand of plywood industries in the country and the systematic introduction of exotic poplars dates back to 7 decades ago. There are over 13 different species representing more than 600 clones and 266 open poplars families were introduced in to India from different parts of the world. Currently it is estimated that poplars are grown in over 3.12 lakhs hectare in the states of Punjab, Haryana, Uttar Pradesh, Himachal Pradesh, Uttranchal, and Jammu and Kashmir. The wood obtained from poplars is an excellent raw material for plywood, particle board, match splints and paper industries. The farmers are more attracted towards poplars cultivation. This exotic species contributes more towards socio economic status of tree growing farmers in several north Indian states. Among various poplar species, the potential of *Populus deltoides* is very well witnessed. WIMCO and ICFRE are the major organizations responsible for the introduction and promotion of poplars in India.

2. Eucalyptus

Eucalyptus was introduced into India in the year 1790 by the then ruler of Mysore "Tipu sultan". It is known that 16 different species have been introduced for ornamental purposes. This was followed by introduction and extensive plantation established in the year 1843 in the Himalayan slopes and also in Nilgiri hills of

Western Ghats. Eucalyptus has been successfully grown almost across the country except north eastern region. Over 70 different eucalyptus species have been introduced in India and two species like *Eucalyptus tereticornis* and *Eucalyptus camaldulensis* are more dominant in the plains and *Eucalyptus grandis* and *Eucalyptus globulus* are introduced in higher altitudes. Currently eucalyptus is extensively cultivated in the states of Tamil Nadu, Andhra Pradesh, Karnataka, Punjab, Haryana, Uttar Pradesh, Madhya Pradesh, parts of Maharashtra and Gujarat. Due to its multifarious industrial utility, all the paper industries in the country have grown extensively eucalyptus plantations as a source of pulpwood raw material. Similarly all plywood industries in the country deployed Eucalyptus as a major raw material for composite wood production. Currently, India accounts for over 0.5 million ha of eucalyptus plantation to cater to the needs of raw material requirement for both domestic and industrial utility. One of the leading paper industries in the country , M/S ITC Ltd company have released several clones in Eucalyptus both through selection and hybridization programme and these clones are now found intensive cultivation across the country. TNAU and ICFRE institutes also released superior clones in Eucalyptus for varied utility.

3. *Casuarina*

Casuarinas are versatile group of plants introduced from Australia six decades ago. Several species were introduced into India and among them the following two species viz., *Casuarina equisetifolia* and *Casuarina junghuhniana* are more prominent. Originally most casuarinas are grown only along the coastal lands and adjacent land use systems: but currently casuarina is cultivated more intensively in the inland conditions. Casuarinas are fast growing species with wide ranging adaptability and grown in different environment and edapho climatic conditions. Casuarinas are recognized as multipurpose tree species and provide multiple end users and services, The wood of casuarina is an excellent raw material for pulp and paper industries. The casuarina poles are extensively used in construction industry as scaffolding materials. The wood of casuarina has high colorific value and hence used as a domestic and industrial fuel. This species fixes atmosphere nitrogen through symbiotic association with the micro organism called Frankia. Currently casuarina is grown extensively in the states of Tamil Nadu, Karnataka, Andhra Pradesh, Maharashtra, Orissa, Gujarat and Puducherry. It is estimated that casaurina is cultivated at an area of over 5 lakhs hectare and contributes significantly to the casaurina growers. TNAU and IFGTB have released wide range of clones in this species and are now under commercial cultivation through agro and farm forestry systems.

4. Subabul (Leucaena leucocephala)

Subabul is pronounced as a miracle tree due to its multiple utility like firewood, timber, fodder, green manure, and as an excellent species for soil and water conservation. The species has been introduced into India during 1970s from Hawaii. Several accessions were introduced in the country but among them the potential of two accessions viz., K8 and K636 is very well witnessed. Subabul is the leguminous tree and with the symbiotic association with rhizobium it is able to fix 500 kg Nitrogen/ha/annum and contributes significantly to the soil amelioration and augmentation. The wood of subabul is medium density and recorded excellent pulp yield coupled with acceptable kappa number and hence the species is used as a raw material by pulp and paper industries. The wood of subabul is a medium textured, close grained and easily workable for a wide range of carpentry purposes like sawn timber, mine props, furniture, and flooring. It has a good calorific value and hence used as a raw material for biomass power industry. The species is currently promoted extensively in the states of Andhra Pradesh, Gujarat, Maharashtra and Tamil Nadu. TNAU has developed an advanced variety in subabul called FCRI LL15 which is a fast grower and has a potential productivity of 150 tones/ha of biomass in 30 to 36 months.

5. Prosopis (*Prosopis juliflora*)

Prosopis juliflora is commonly called as Mesquite has been introduced into India in the year 1870 for afforestation purposes from Central and South American countries. The species is a fast growing and identified as a rapid colonizer. It is a multipurpose tree and grows very well in all types of soil and environmental conditions. The wood of prosopis has recorded high calorific value of over 5000 Kcal/kg and hence meets the domestic firewood requirement of the country. Over 60% of the country's energy demand is met from *Prosopis juliflora*. The pods of *Prosopis juliflora* are rich source of protein and hence act as an excellent feed supplement for the cattle. It is a profused seed bearer and hence easily colonizes the new locality. Because of this character, the species is considered as an exotic invasive species and received wide range of criticism among the environmentalists. However it is a fast grower, stabilizes soil, control erosion, improve soil fertility and provide multifarious end uses. The species is highly amenable for sodic and alkaline soil and plays a vital role in reclamation of problem soils.

6. Ratanjothi (*Jatropha curcas*)

Jatropha curcas is a central American species introduced during 16th century by the Portuguese settlers. The species has been introduced as a fence and ornamental crop and found rapid cultivation across the country as a source of live fence. However, due to changes in policy directions, the government of India has identified *Jatropha curcas* as the potential oil seed crop which could be used for biofuel production. Based on these policy directions, *Jatropha curcas* has been introduced and cultivated extensively across the country. The species yields an oil to the tune of upto 30% and upon transesterification, the oil yields biodiesel (Methyl ester). The oil cake is also found utility as an organic manure with high nitrogen content. The fruit shells are used as a fuel. The species is found amenable for wide range of environmental conditions and large scale cultivation has been attempted in the states of Tamil Nadu, Karnataka, Andhra Pradesh, Maharashtra, Uttar Pradesh, Madhya Pradesh, Gujarat, Rajasthan, Chhattisgarh etc., TNAU has developed high yielding hybrids which are now under various stages of growth and development.

7. Silver Oak – *Grevillea robusta*

Silver oak is a fast growing multipurpose tree species introduced from Australia as an ornamental shade giving tree. The species is highly amenable for both tropical and sub tropical climate with an altitude ranging 300 m to 1500 m above MSL. It is one of the best and ideal species as a shade tree for coffee and tea plantations. Due to its spreading branching character, the species is found amenable for windbreak or shelterbelts and protect the wind induced mechanical damage and reduces the moisture loss due to evaporation. It is a medium density wood and the density varied between 540 kg/m^3 and 720 kg/m^3. The wood is used in construction industry, furniture and plywood industries. Above all, it is an excellent landscape tree and planted in the public and private gardens and landscapes. The species is grown as a shade crop for tea and coffee plantations in the states of Tamil Nadu, Karnataka, Kerala and several Northern eastern states.

ADVANTAGES OF EXOTIC SPECIES

- In the tropics and subtropics, the exotic coniferous timbers have a great need to their superior wood quality because the indigenous forests frequently do not contain the required quality wood.
- The productivity of indigenous forests are very low of 0.5 m^3/ ha/ annum but the exotic plantations have the production potential of 25-30 m^3/ ha/ annum and hence they are preferred predominantly for productive industrial wood plantations.
- Exotics have wider adaptability and multipurpose utility compared to the indigenous species which are attracted by several forest based industries.
- Most of the exotics are fast growers which results in lowering of the rotation period and increased economic returns. Some of the exotics like *Casuarina* and *Eucalyptus* are harvested in less than 5 years wherein no indigenous species are harvested within this period.
- The tropical and indigenous hard wood forests are so variable and are difficult to manage silviculturally than the exotics. On the other hand, exotics are easy to manage both silviculturally and ecologically due to their short rotation, fast growing and multiple utility. Hence exotic are preferred in place of native species.
- Most foresters generally prefer to use species that are more uniform and whose products are known and accepted. Hence the exotics are widely accepted as a species with uniformity and easy to manage.

- Marketing potential and manufacturing technology favors the use of exotic woods such as Pines, Poplars, *Eucalyptus* etc., compared to indigenous species.
- Availability of quality seed and planting stock is a key to success in any plantation forestry programme. The seeds of the indigenous species are sometimes hard to obtain due to poor understanding of reproductive biology. On the other hand, several exotic species have been developed into improved genetic stock resulting in better quality and more uniform trees from which seeds are easily available.

PROBLEM AND CONSTRAINTS DUE TO EXOTICS

- Immediate failure of the introduced plantations is one of the major problems of the exotics.
- There are instances of delayed failure of exotics. Initially the exotics species exhibit good survival and growth but over periods they do not develop into useful forests.
- The exotics may have a continued substandard performance resulting in low production and poor quality.
- The exotics have initial survival and growth but later on attacked by pest and diseases resulting in complete failure of the plantations.
- Growth of the exotics may be unsatisfactory due to shortage or absence of mycorrhizae.
- Adverse weather conditions like drought, cold and violent weather are the causes for failure of exotics. Extremes of weather result in loss in growth and quality and sometimes lead to the death of the plants.
- The newly established exotics may be initially pest free but later on it is infested with pest and diseases resulting in complete failure of the plantations.
- In certain cases, exotics also one of the causal factor for poisoning of the soil through allelopathy effect due to the secretion of toxic substances.
- Few exotics develop abnormality in its new environment resulting in loss of apical dominance, crooked development and with multiple tops.

- Exotics also can change the quality and amount of water supply that drains from the forests.
- Some of the exotics like eucalyptus and prosopis have been debated towards their role in depletion water table and is considered as a major threat by the ecologist and environmentalist.

SUMMARY

Exotic species played a very vital and significant role in meeting fuel wood demand besides satisfying the raw material needs of several wood based industries. Among several wood based industries, the plywood sector and paper industries heavily dependent on exotic species like Eucalyptus, Poplars, Casuarina and Subabul. The role of eucalyptus in satisfying the raw material needs of plywood industries in the country is very well witnessed. This exotic species also contribute significantly in the growth of ply and veneer industries in the country. Similarly the use of eucalyptus in paper industry for different grades of paper production is also very well established. The other exotics like Casuarina and Subabul in South India and poplars in north India are potential exotics which are used as a raw material in several wood based industries. Country's like India, wherein the wood supply from natural forest is restricted and demand extensive establishment of plantations using fast growing exotics. Under such circumstances, over 20 different exotic species amenable for both tropical and temperate climates are popularized and promoted in the country in various afforestation, reforestation and agroforestry programmes. These exotics have exhibited several advantages like easy adoptability, fast growth, higher biomass production, adoption to wide range of edapo-climatic condition and suitable as a raw material for wide range of industries. However, the introduction of exotics has also created criticism for the negative impacts of exotics on the ecology and ecosystem of the country. All facets of the exotics are furnished in the chapter which will be a great benefit to the readers of this book.

14

TISSUE CULTURE AND ITS ROLE IN FORESTRY

INTRODUCTION

Tree breeding and improvement programme has received significant strength due to the incorporation of tissue culture technology towards exploiting the superior genetic stocks developed through traditional tree breeding programme. Predominantly most tree species which are commercially important such as timber, plywood, pulpwood, bio fuel, match splints, sports goods, agricultural implements etc., exhibited high degree of heterozygosity in the natural population. The high degree of heterozygosity demanded technologies for mass multiplication of potential genetic resources. Similarly even the controlled breeding populations that resulted in F_1 generation due to the crosses between widely variable parents leads to highly variable F_1 generation. This also needs a technology towards exploitation of hybrid vigour through mass multiplication. Under such circumstances, plant tissue culture technology has been identified as one of the potential and robust technologies for a large scale multiplication of wide range of tree species. The science of plant tissue culture has been very well exploited and technologies have been standardized for a wide range of commercially important tree species. Hence, this chapter is designed and incorporated the basics and fundamentals of tissue culture technology along with its application in commercially important tree species towards large scale multiplication and conservation.

HISTORY OF PLANT TISSUE CULTURE

The plant tissue culture has its origin from the discovery of cell theory which was visualized by Schleiden and Schwan (1983) and indicated that cell is the basic structural unit of all living organism and has a capability of regeneration into a whole plant. This basic idea and concept was further developed by a German physiologist Gottlieb Haberlandt who attempted to culture cells and proposed the theory of totipotency i.e., ability of a plant cell to develop into a whole organism. Due to this, Haberlandt has been named as Father of plant tissue culture. This was followed by wide range of invention and this chapter traces history and development of various achievements in the science of plant tissue culture.

S.No	Year	Author	Development
1.	1902	Haberlandt	Proposed the theory of in vitro cell active through his experiment in isolation and culturing of cells on knobs salt solution. He is named as Father of tissue culture.
2.	1904	Hannig	Cultured embryos of cruciferous species and got successful growth.
3.	1922	Kolte & Robbins	Successful culturing of root and stem tips.
4.	1926	Went	Discovery of first plant growth harmone Indoleacetic acid (IAA).
5.	1925–29	Laibach	Zygote embryo culture was demonstrated practically.
6.	1934	White	Vitamin B was used as a growth supplement in tissue culture media for the culturing of tomato root tip
7.	1939	Gautheret White Nobecourt	Established large scale multiplication and proliferation of callus culture.
8.	1941	Overbeek	First time deployed coconut milk for cell division in Datura.
9.	1946	Ball	Developed shoot tip culture techniques for plantlet development in Lupinus species.

S.No	Year	Author	Development
10.	1954	Muir	He was the first scientist to break callus tissues into single cells and attributable to single cell culture.
11.	1955	Skoog & Miller	Discovered kinetin as a cell division hormone.
12.	1957	Skoog & Miller	Developed the concept of hormone control (Auxin/ Cytokinin) of organ formation.
13.	1959	Rineet& Steward	Embryo regeneration from callus and cell suspension culture of carrot
14.	1960	Cocking	First isolation of protoplast by enzymatic degradation of cell wall.
15.	1960	Morel	Developed of virus free plants by shoot tip culture and attributable to meristem culture.
16.	1960	Bergmann	Single cell isolation by plating through filtration of cell suspension culture.
17.	1960	Kanta & Mageswari	Developed first Test tube fertilization technique.
18.	1962	Murashige & Skoog	Developed MS medium with higher salt concentration. This medium is now recognized as a standardized basal medium for protocol development in various plant and tree species.
19.	1964	Guga & Mageswari	Produced first haploid plants from pollen grains of *Datura metal* through anther culture.
20.	1970	Power & his Co-worker	Successful fusion of protoplast.
21.	1971	Nagata & Takebe	First regeneration of plantlets from protoplast.
22.	1972	Carlson	First interspecific hybrid development in Tobacco through protoplast fusion - first regeneration of somatic hybrid.
23	1974	Reinhard	Introduced bio transformation in plant tissue culture.

S.No	Year	Author	Development
24	1977	Chilton & his Co workers	Successful integration of Ti-plasmid from *Agrobacterium tumefaciens* in plants.
25	1978	Melchers & his Co workers	Somatic hybridisation of tomato and potato resulting in development of pomato.
26	1976	Scibert	Development of cryo preservation techniques.
27	1981	Larkin & Scowcroft	Introduction of somo clonal variation.
28	1982	Zimmerman	Protoplast fusion through electro-fusion.
29	1983	Pelletier & his Co workers	Studies on inter – generic cytoplasmic hybridisation in radish and grape.
30	1984	Horsh & his Co workers	Developed transgenic tobacco through Agrobacterium mediated transfer.
31	1987	Klien & his Co workers	Developed biolistic gene transfer (particle bombardment) for plant gene transfer.
32	1987 to 2018	Wide range of authors	Developed and reported tissue culture protocol, meristem culture, anther culture, protoplast culture, protoplast fusion, cell suspension culture, somatic embryos and gene transfer techniques in wide range of annuals and perennials.

DEFINITION OF TERMINOLOGIES

Tissue culture : *In vitro* culture of cells, tissues, organs and plants under aseptic conditions on synthetic media.

Totipotency : Capacity of plant cells to regenerate whole plants when cultured on appropriate media.

Callus : Callus is an unorganized mass of differentiated plant cells.

Cell culture	:	Culture of cells or their maintenance *in vitro* including the culture of single cells is Cell culture.
Culture	:	A plant growing *in vitro* in a sterile environment
Embryo culture	:	*In vitro* culture of isolated mature or immature embryos.
Explant	:	Explants are an excised piece or part of a plant used to initiate a tissue culture.
In Vitro	:	To be grown in under controlled condition.
In Vivo	:	To be grown naturally
Micropropagation	:	Multiplication of plants from vegetative parts by using tissue culture nutrient medium is Micropropagation
Propagule	:	A portion of an organism (shoot, leaf, callus, etc.) used for propagation.
Subculture	:	The aseptic division and transfer of a culture or portion of that culture to a fresh synthetic media.
Morphogenesis	:	Morphogenesis is the biological process that causes an organism to develop its shape.
Organogenesis	:	Formation of organs from the cells.
Embryogenesis	:	Embryogenesis is defined by a sequential series of dynamic processes that include cell division and growth, and the elaboration of differentiation programs leading to cell fate specification.
Callogenesis	:	Callogenesis is the process of callus formation.
Caulogenesis	:	Caulogenesis is the type of organogenesis by which only adventitious shoot bud initiation take place in the callus tissue.
Rhizogenesis	:	Rhizogenesis is the type of organogenesis by which only adventitious root formation takes place in the callus tissues
Plant Growth Regulators	:	Plant growth regulators may be defined as any organic compounds, which are active at low

concentrations (1-10 ng / nl) in promoting, inhibiting or modifying growth and development. The naturally occurring (endogenous) growth substances are commonly known as plant hormones, while the synthetic ones are called growth regulator.

PLANT TISSUE CULTURE

It is the science of growing plant cells, tissues or organs isolated or separated from the mother plant or artificial growing medium under *in-vitro* conditions. Tissue culture and micro propagation are predominantly used terminologies to describe all kinds of *in-vitro* plant cultures. However, these two terminologies strictly refer to cultures of organized aggregates of plant cells (Micro propagation) and unorganized aggregates of cells (Tissue culture). The micro propagation strictly deals with direct organogenesis without callus mediated process. Tissue culture involves a callus mediated process. The micro propagation may result in true to type and ensures clonal fidelity. But the tissue culture may or may not be resulting in somaclonal variants. However, for all practical purposes both the terminologies of tissue culture and micro propagation are used to describe all kinds and methods of in-vitro plant cultures to benefit the readers of the book.

ADVANTAGES AND SIGNIFICANCE OF PLANT TISSUE CULTURE

- Tissue culture helps to mass multiply true to type Forest Genetic Resources and will maintain the clonal fedility.
- The tissue culture techniques help to produce disease free plants through the meristem culture.
- Tissue culture plays a role in production of homozygous plants through anther culture.
- The development of somatic hybrids through protoplast fusion is also possible through plant tissue culture technology.
- Development of synthetic and artificial seeds is another major contribution by tissue culture.
- Tissue culture also helps in production of secondary plant metabolites through cell suspension culture.

- Tissue culture technology plays a vital role in conservation of forest genetic resources through cryopreservation technology.
- Development of sterile varieties is another major contribution by tissue culture technology through endosperm culture.
- Tissue culture technology helps in exchange of genetic resources both nationally and internationally due to its *in-vitro* nature and free from pest and diseases.
- It is a significant technology for those species which are recalcitrant towards multiplication through seeds and for those species which are sterile.
- Rare, threatened and endangered species can very well be mass multiplied through tissue culture technology and will help in regeneration, and rehabilitation of RET species.

TYPES AND METHODS OF TISSUE CULTURE

The tissue culture based on growth and development and the nature of tissues involved are grouped into two major types:

A. Organized tissue culture

B. Unorganized tissue culture

A. ORGANIZED TISSUE CULTURE

It is a type of culture wherein an organized form of growth of tissues is continuously maintained. In this type, the following organ based cultures are carried out towards multiplication of organs.

i. Shoot tip culture

It is a type of organized tissue culture wherein the shoot tips or shoot buds with several leaf primordia are excised from identified mother plant and deployed in in-vitro culture.

ii. Meristem culture

It is a type of organ culture wherein the shoot tips containing apical meristematic dome with or without leaf primordia are excised and deployed in in-vitro culture.

iii. Nodal culture

It is a type of organ culture wherein the nodal segments or lateral buds or stems with single or multiple nodes are separated from the identified mother plant and deployed in culture under *in-vitro* conditions.

iv. Root tip culture

Some of the species are highly responsible for root organ culture wherein the root tips of plants are excised and deployed for *in-vitro* growth and development.

v. Embryo culture

It is a type of organ culture wherein the fertilized or unfertilized zygotic embryos are dissected out from the developing or developed seeds and are cultured under *in-vitro* conditions.

2. UNORGANIZED TISSUE CULTURE

i. Callus culture

It is a type of unorganized tissue culture wherein, callus - undifferentiated mass of parenchyma tissues are induced from an explants under *in-vitro* conditions.

ii. Suspension culture

The unorganized callus cultures grown in liquid or cell suspension medium under agitated and aerated conditions towards cell growth and differentiation under in-vitro conditions are termed as cell suspension cultures.

iii. Protoplast culture

It is a type of unorganized tissue culture wherein the plant cells without a cell wall is cultured under *in-vitro* conditions.

iv. Anther culture

Culturing of anthers containing pollen microspores under *in-vitro* conditions leading to the development of haploid plants is termed as Anther culture.

TISSUE CULTURE LAB AND REQUIREMENTS

For establishment of tissue culture or micro propagation, there is a need to have a well designed tissue culture lab with all the essential facilities and equipments. The following laboratory facilities are required to carry out propagation studies for various tree species. The pilot structure of an institutional tissue culture lab is designed and furnished below.

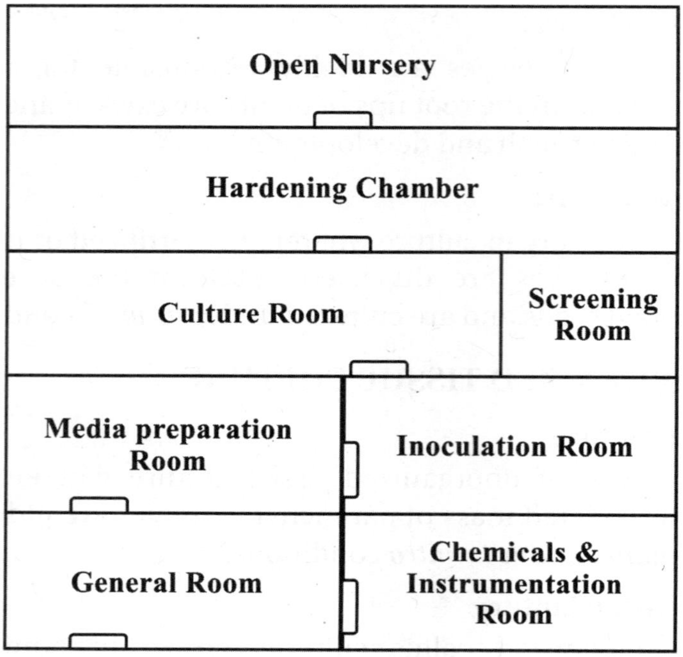

Fig 14.1. Pilot Tissue Culture laboratory Designed for Academic / Research Institution

Laboratory area- 500 sq. ft

Features of the lab

- General lab facility
- Chemical room
- Media preparation and storage room
- Inoculation room
- Culture room
- Screening room
- Hardening chamber

Inoculation

Culture room

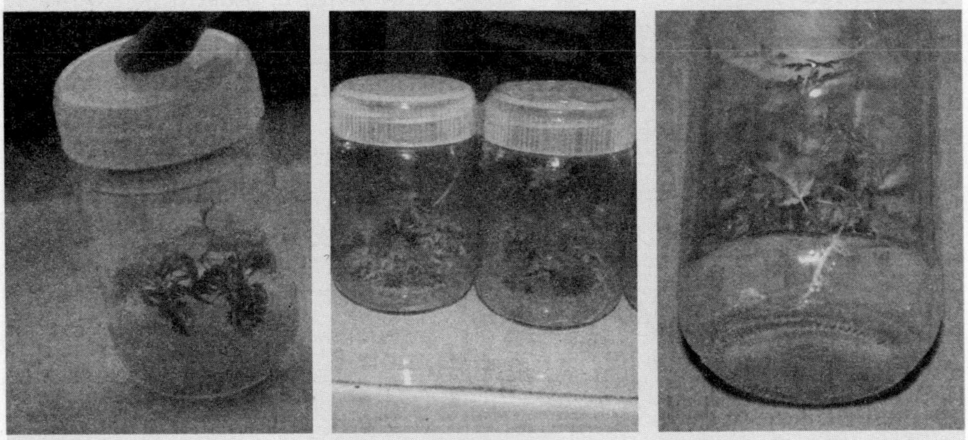

Established plants

Fig 14.2. TISSUE CULTURE LAB AT FCRI, METTUPALAYAM

Equipments and other laboratory requirements for a pilot lab

S. No.	Equipment	Brief Specs.	Nos.
1	Laminar Air Flow Chamber	6, 2 seater, SS	1
2	Glass Bead Sterilizer	Auto temp/timer	2
3	Working chair	Ergonomic	4
4	Autoclave 1	Automatic, 300 bottle	1
5	Autoclave 2	Automatic, 150 bottle	1
6	Balance	Digital, 0.001 mg	1
7	Magnetic stirrer	With hot plate, 5 lit	1
8	Heat sealer	9``	1
9	Media stirrer	With stand	1
10	Thermo hygrometer	Digital	1
11	Growth rack	With light LED, timer	1
12	Pass box	With UV lamp	1
13	Pressure module		1
14	Trays	Autoclavable, plastic	300
15	Bottles	300 ml	10000
16	Caps	Autoclavable, plastic	10000
17	Induction stove	1200 W	1
18	Refrigerator	220 L	1
19	Bottle washer	Optional	1
20	Air conditioners	1.5 ton	2
21	Work table media	-	1
22	Work table office	Optional	2
23	Work stools	SS top	3
24	Fire Extinguishers	Standard	2
25	Lab accessories	-	-
26	Chemicals	Macro, micro, vitamin, growth regulators, Agar, sucrose and other essential chemicals	Required quantity

VARIOUS STAGES OF TISSUE CULTURE

The sequential stages recognized in any Tissue culture / micro propagation systems involved are

i) Establishment or Initiation Stage
ii) Multiplication or shoot development stage
iii) Pre- transplant or Rhizogenic stage
iv) Transplant or hardening Stage

i. Establishment stage / Initiation Stage

The factors that affect establishment stage are

i) Choice of explants
ii) Elimination of contamination of the explant and
iii) Culture conditions

In general, younger tissue such as shoot tips or terminals buds will regenerate better than older and mature tissues of the same stem. In general, ingredients of the culture medium in the first stage are determined by kind of response needed. The culture room conditions involve a temperature range of 25 + 2°C and with a photoperiod of 16 hrs. light and 8 hrs, darkness. This stage last for 4 to 8 weeks.

I. EXPLANTS

The organs or tissues that are excised from identified mother plants diploid for *in-vitro* culturing is termed as explants. There are wide ranges of explants like shoot tips, meristems, nodes, root tips, flowers, seeds, anthers, pollen etc., are deployed in tissue culture technology.

II. STERILIZATION

The identified organs or the plant material normally carries a range of micro-organisms and hence the explants need to be sterilized using chemicals and in the specially designed rooms. The following sterilants are predominantly used.

S.	Major Chemicals	Concentration (%)	Duration (minutes)
1	Mercuric chloride	0.1-1.0	2-10
2	Sodium hypochlorite	0.5-5	5-30

S.	Major Chemicals	Concentration (%)	Duration (minutes)
3	Calcium hypochlorite	9-10	5-30
4	Hydrogen peroxide	3-12	5-15
5	Ethyl alcohol	70-95	0.1-5
6	Silver nitrate	1	5-30

The identified explants are separated from the mother plants collected in a conical flask containing distilled water and washed thoroughly. Then few drops of soap solution or teepol are added with water and washed thoroughly which eliminates the surface organisms and facilitates fusion of sterilants inside the explants.

After washing in soap solution, the explants are transferred to sterilized hood chamber normally laminar air flow chamber is used in most laboratories. This laminar air flow chamber filters atmospheric air and provides micro-organism free air. Under this condition, the explants are treated with suitable sterilant with standardized concentration and duration followed by continuous washings for 3-5 times using sterile distilled water. The explants are now ready for culturing in an artificial medium.

III. MEDIA

The medium used for tissue culture / micro propagation has two major functions.

i. To supply basic ingredients for continued growth of the isolated explant and subsequent propagation and

ii. To direct growth and development through hormonal control

The artificial medium containing major, minor, vitamins and growth regulators along with solidifying agent (Agar) and energy source (Sucrose) used in culturing of plant tissues under *in-vitro* condition is called as media. Various authors and scientists have developed a wide range of media composition for different species but the most commonly used media for growth and development of tree species are furnished below

i) MS medium - Murashige and Skoog, (1962)

ii) Woody plant medium - Lloyd and McCown, (1981)

iii) N6 medium - Nitsch & Nitsch, (1969)

iv) B5 medium - Gamborg, (1968)

MEDIA PREPARATION

Media play a vital role in development of plant organ in micropropagation or tissue culture process. Similar to the seedling growth and development, tissue culture plants also demands wide range of nutritional requirement during the stages of initiation, multiplication followed by growth and development into organs. The explants intended for micropropagation will grow only under *in-vitro* conditions provided with a specific media composition. This media composition consists of wide range of nutritional supplements which are indicated below.

i. Macro Elements

These macro elements are required in larger quantities for initiation of plant materials followed by its subsequent growth and development. The major nutritional macro elements are Nitrogen, Phosphorus, Potassium, Calcium, Magnesium and Sulphur.

ii. Micro Elements

These micro elements are also very important to catylyze several plant biosynthetic process and promote plant growth and enzymatic system. Some of the micro element are actively involved in cell division and growth and helps to complete the plant growth and development process. The major micro elements required are Boron, Copper, Zinc, Iron, Manganese and Molybdenum.

iii. Organic supplements

The organic supplement includes both vitamins and amino acids. The vitamins particularly thiamine and myoinositol are essential for the culture of the plant cells for *in-vitro* cultures. The amino acids are rarely used in tissue culture but depend on the species and as a source of reduced Nitrogen, it is incorporated based on the need.

iv. Plant Growth Regulators

These are small organic molecules involved in various physiological process and act as a facilitator between the environment and genome. These growth regulators are synthesized naturally by plants but their exogenic application received potential attraction for organogenesis and Rhizogenesis. The hormonal is control exercised by the following factor

i. Kind of hormone or growth regulators

ii. Its concentration and

iii. Sequence in which they are applied.

The major plant growth regulators used in Plant Tissue Culture are

a. Auxins

b. Cytokinins

c. Gibberellins

d. Abscisic acid

e. Ethylene

Among these plant growth regulators, the role of auxins and cytokinins is very well witnessed for successful development of both shoots and root organs. The higher auxin to cytokinin ratio leads to more of root development and higher cytokinin to auxin ratio leads to shoot development and hence their utility in plant tissue culture is most significant.

v. Antibiotics

These are substances produced by certain micro organisms that suppress the growth and development of other micro organism and ultimately destroy them. During the process of incubation, the culture media and plant material expressed contamination due to micro organisms and affect the *in-vitro* conditions of the plant irrespective of the sterilization process. Under such circumstances, use of antibiotics like chloramphenical, tetracycline, rifampicin, streptomycin, neomycin etc., received significant attention in plant tissue culture.

vi. Solidifying Agents

In most plant tissue culture, growth and development are carried out in a semisolid gaseous medium which provides support along with required gaseous exchange under static conditions. The most common gelling agents used is agar followed by gelrite.

vii. Energy Source

The predominant energy source used in most culture media is sucrose. The composition of various nutritional media developed and used as a basal media composition in tissue culture protocol development of various tree species are furnished in following tables.

S. No	Components	MS Medium	Woody Plant Medium	B5 Medium	Nitsch Medium
a. Macrosalts					
1.	KNO_3	1900	-	2500	950
2.	NH_4NO_3	1650	400	-	720
3.	$MgSO_4.7H_2O$	370	180.54	246	90.27
4.	$CaCl_22H_2O$	440	75.5	150	166
5.	KH_2PO_4	170	170	-	68
6.	$Ca(NO_3)_2$	-	386.8	-	-
7.	K_2SO_4	-	990	-	-
8.	$(NH_4)_2SO_4$	-	-	134	-
9.	$NaH_2PO_4.H_2O$	-	-	150	-
b. Microsalts					
10.	H_3BO_3	6.2	6.25	300	10
11.	$MnSO_4.4H_2O$	22.3	22.3	10	25
12.	$ZnSO_4.7H_2O$	8.6	8.6	2	10
13.	$Na_2MoO_4.2H_2O$	0.25	0.25	0.25	0.25
14.	$CuSO_4.5H_2O$	0.025	0.25	0.025	0.025
15.	$CoCl_2.6H_2O$	0.025	-	0.025	-
16.	KI	0.83	-	0.75	-
c. Vitamins					
17.	Nicotinic acid	0.5	0.5	0.5	0.5
18.	Pyridoxine HCl	0.5	0.5	0.1	0.5
19.	Thiamine HCl	1	1	0.5	0.5
20.	Myoinositol	100	100	100	100
21.	Aminoacid-glycine	2	2	2	2

22.	Folic acid	-	-	0.5	-
23.	Biotin	-	-	0.05	-
d. Iron EDTA					
24.	$FeSO_4.7H_2O$	27.8	-	27.8	27.8
25.	Na_2 EDTA	37.3	-	37.3	37.3
26.	Fe_2Na EDTA	-	36.7	-	-

PREPARATION OF STOCK SOLUTION

Preparation of stock solution for various nutritional ingredients is very essential because the culture medium requires a minimum of 17 to a maximum of 25 different chemicals. Weighing every time all the ingredients of a media is a time consuming process and also demand skilled man power. Hence for easy and convenience and to save the time and man power, it is very essential to prepare and keep the concentrated stocks of macro elements, micro nutrients, vitamins, plant growth regulators, amino acids etc., in the refrigerator conditions at -4 to +4 degree temperature. The stock solution preparation for various mineral elements is furnished below taking the examples of Murashige and Skoog's basal medium.

i. Stock solution for macro nutrients

To prepare stock solution for macro elements, the final concentration in the medium should have to be multiplied by 10 times. The 10 times multiplied mineral nutrients are dissolved in 500 ml double distilled water by dissolving each mineral constituent separately one by one. This 500 ml will act as a stock solution for macro elements and from this 5 ml of the stock solution can be taken out to prepare 1 litre medium.

ii. Minor nutrients

The minor nutrients particularly manganese, zinc, boric acid can be prepared at 100 times of their final concentration. The minor nutrients are weighed individually and dissolved separately with double distilled water and final volume is made up to 100 ml. This 100 ml stock solution is enough to prepare 100 litres of medium by taking 1 ml from the stock solution to prepare 1 litre of the medium.

iii. Iron stock

The iron- EDTA stock solution has to be prepared separately by mixing them 50 times of their final concentration as indicated below

- 1.393 g of $FeSO_4.7H_2O$ is dissolved in 100 ml of water and the solution is gently heated.
- Similarly 1.863 g of Na_2-EDTA is dissolved in another container containing 100 ml of water and is gently heated.
- The warmed solution of EDTA has to be added to the stock solution of $FeSO_4$ and the final volume is made up to 250 ml which will act as a stock solution. This stock solution looks like deep golden yellow in colour and stored in amber colour bottle in the refrigerator.

iv. Potassium iodide stock solution

The potassium iodide stock solution has to be prepared separately by multiplying 100 times of its final concentration and dissolved in double distilled water and the final volume is made in to 100 ml. From this stock solution, 1ml can be taken to prepare 1 litre of the medium.

v. Vitamin stock solution

The vitamins are weighed multiplying 100 times of their final concentration and dissolved separately in double distilled water. The final volume is made to 100 ml which is amenable for preparing 100 litre of medium i.e., 1ml of the stock solution is enough to prepare 1 litre of the medium.

vi. Stock solution for growth regulators

It is better to always prepare freshly the plant growth regulator solutions. To prepare stock solution, it is better to individually prepare separate stock solution for respective auxins and cytokinins. Generally, the auxins and cytokinins do not dissolve in distilled water and hence these growth regulators have to be dissolved in necessary solvents.

Auxins: Auxins are to be dissolved in 2.5 ml of 95% ethanol or 2.5 ml 1 M KOH or 1 M Sodium hydroxide through slight heating. Then the final volume is made in to the required volume. For example, if 100 mg of auxin is dissolved in 2.5 ml of ethanol it is finally made into 100 ml stock solution. If 1 ml of the stock solution is taken, it indicates 1 mg. Similarly if 5 ml of stock solution is taken it indicates 5 mg. Accordingly, the auxin and cytokinin can be prepared and added to the media.

Final steps for media preparation

i. Necessary glassware preferably 2 litre volumetric flask have to be taken.

ii. The following quantities of stock solutions are to be taken out from the already prepared and stored stock solution.

Stock 1 - 50 ml

Stock 2 - 1 ml

Stock 3 - 1 ml

Stock 4 - 1 ml

Stock 5 - 1 ml

iii. After adding all stock solution, the energy source namely 30 g of sucrose is added and dissolved in the medium.

iv. Now the required quantities of growth regulator- auxin /cytokinin are to be added.

v. Finally the entire media is to be adjusted to near volume using double distilled water.

vi. The entire media is now checked for pH between 5.6 ± 0.1 by adding with either 0.2 N Sodium hydroxide or 0.5 N HCl with necessary stirring using glass rod.

vii. Now the final volume of media should be adjusted to 1000 ml using double distilled water.

viii. Then the media is taken to microwave oven wherein required quantity of agar (6-9 g/l) or gelrite (2-4 g) is added by slow release followed by stirring to facilitate uniform melting of the gelling agent. It can be done in the Bunsen burner also.

ix. After complete melting of agar, the media is distributed into the test tubes or tissue culture bottles with required volume.

x. Then the media with containers are to be autoclaved at 15 lbs at 121°C for 20 minutes.

xi. After autoclaving is over, the media is labelled properly and stored in sterile laboratory conditions preferably at culture room conditions.

Preparation of Stock Solutions for MS medium

Ingredients	Composition (mg^{-1})	Stock Solution (W/V) (g)		Volume of the stock taken / litre of medium
Macro Nutrients (10X)				
NH_4NO_3	1650	16.50		
KNO_3	1900	19.00		
$CaCl_22H_2O$	440	4.40	500 ml	50 ml
$MgSO_4.7H_2O$	370	3.70		
KH_2PO_4	170	1.70		
Minor Nutrients (100X)				
$MnSO_4.4H_2O$	22.3	2.23		
$ZnSO_4.4H_2O$	8.6	0.86	100 ml	1.0 ml
H_3BO_3	6.2	0.62		
Micro Nutrients (100X)				
$Na_2MoO_4.2H_2O$	0.25	25		
$CuSO_4.5H_2O$	0.025	2.5	100 ml	1 ml
$CoCl_2.6H_2O$	0.025	2.5		
Iron Stock (50X)				
Na_2EDTA	37.25	1.863		
$FeSO_4.7H_2O$	27.85	1.393	250 ml	5 ml
KI (100X)				
KI	0.83	0.083	100 ml	1.0 ml
MS Vitamins (100X)				
Nicotinic Acid	0.5	50		
Pyridoxine.Hcl	0.5	50		

Ingredients	Composition (mg^{-1})	Stock Solution (W/V) (g)		Volume of the stock taken / litre of medium
Thiamine. HCl	0.1	10	100 ml	1 ml
Glycine	2.0	200		

ii. Multiplication stage / Shoot development Stage

The function of this stage is to increase the number of propagules for later rooting to planting stage. The propagules are cut and separated to be grown into plantlets in a new medium. Number of multiplications may vary from 5 to 50 depending on species and method. Generally, cytokinins are used for multiplication. The major cytokinins used in multiplication stage are :

S.No	Particulars	Chemical Name	Concentration / Litre of medium (Range)
1.	BA	Benzyl Adenine	0.5-5 mg
2.	BAP	Benzylaminopurine	0.1-5 mg
3.	Kinetin	6-Furfurylaminopurine	0.2-5 mg
4.	2iP	2 isopentile adinine	01.-3 mg
5.	Zeatin	(E)-Zeatin Zeatine	0.05-2 mg

In this stage, the multiple shoots initiated in the primary culture are excised separately and are inoculated into the medium containing higher concentrations of cytokinin either singly or in combinations and kept under standard culture room conditions. Due to the presence of basal medium with growth regulators, the shoot buds divide mitotically and induce more number of shoots. The rate of multiplication and the period required could be manipulated by hormonal regulations.

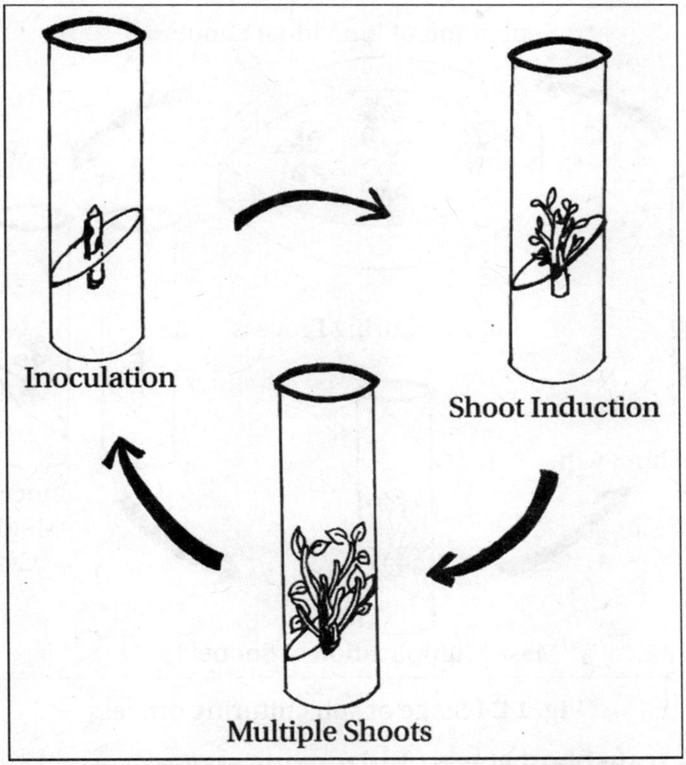

Fig 14.3. Stage of shoot bud multiplication

Sub Culturing

Once sufficient shoot multiplication has occurred, the individual shoot cuttings are dissected out asepitally under laminar flow conditions. These separated shoot cuttings are further inoculated into the same multiplication medium or newly composed subculture medium inorder to induce more shoot buds. These induced shoot buds are further dissected out and cultured on to the same medium. This process of multiplication, separation and inoculation should be continued for 6 to 10 cycles till the morphogenic potential of shoot buds start declining. These shoot buds obtained from series of sub cultures are a base material for the next stage. Depending on the species, the rate of multiplication, period of multiplication, Media composition the number of sub culture cycles may vary.

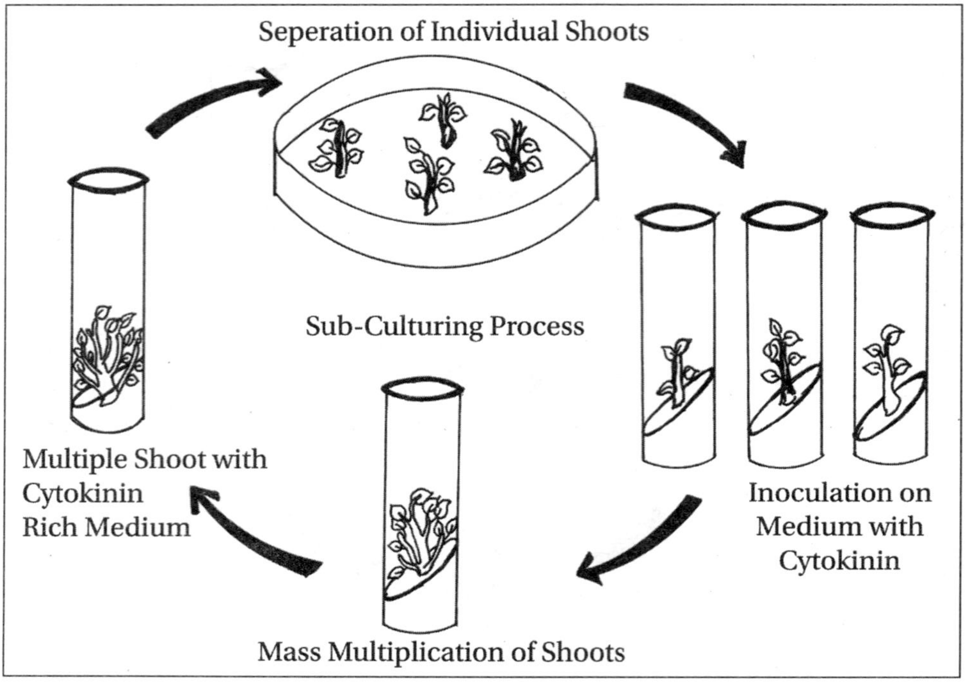

Fig. 14.4 Stage of sub culturing process

iii. Pre – transplant stage / Rhizogenic stage

The function of this stage is to prepare the plantlet for transplanting and establishment outside the artificial, closed or open environment. In this stage, prolific root induction / initiation is the main objective through hormonal regulation. In this stage, the shoot buds obtained from 5 or 6th sub culturing are excised and inoculated into the medium rich in auxins preferably with IAA or IBA at concentrations ranging from 0.5 mg/l to 5 mg/l of medium either singly or in cocktails This is achieved by reducing cytokinins concentration and increasing auxin supply as indicated above. Other conditions required for *in-vitro* culturing with standard cultutre room conditions are to be followed to induce roots from the microshoots. Normally depending on the species, the root induction period varied between 15 days and 90 days for most tree crops. Once rooting is achieved, the *in-vitro* plantlets are need to be exposed to open environment through series of hardening stage.

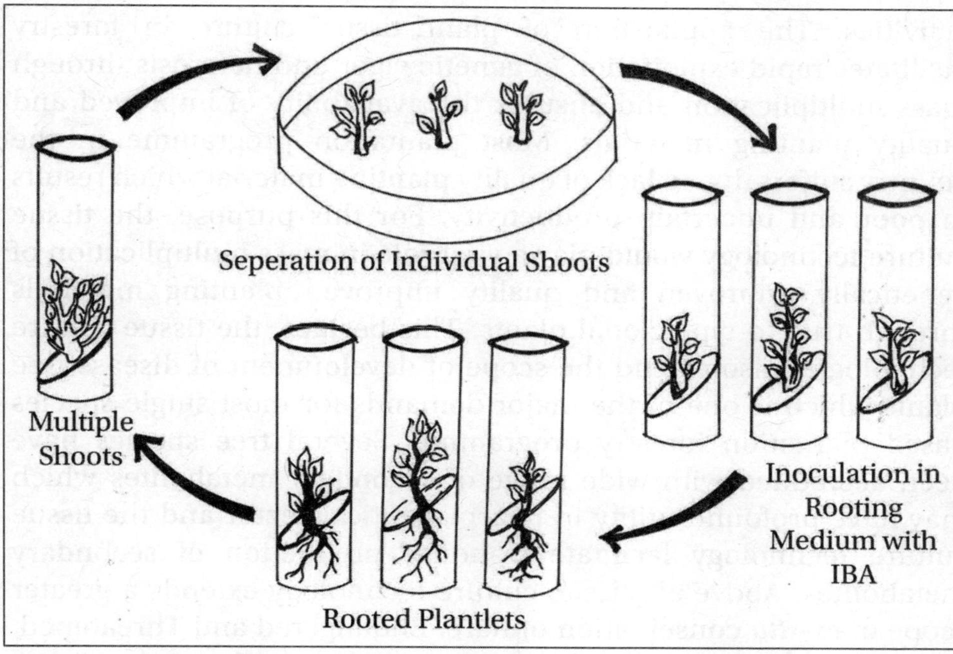

Fig 14.5. Stage of rooting under *in-vitro* condition

iv. Transplant stage

This stage involves the transfer of the plantlet from the laboratory (aseptic cultural environment) followed by free living environment of the green house conditions and ultimately required hardening process to the final field location. The plantlet must not only root adequately but also must undergo a period of acclimatization to enable it to survive under open field conditions. For this purpose, a well designed hardening chamber with 50% or 70 % shade net (With black or Green Colour) house is mandatory to expose partial light to the *in-vitro* cultured plantlets. The infrastructure facilities like misting, fogging, drip and fertigation system needs to be established in the hardening chamber which will facilitate easy and rapid hardening of plantlets. A careful monitoring and management system should be ensured in order to handle the test tube plantlets.

SUMMARY

Application of tissue culture technologies in forestry is one of the most promising areas of present time towards assisting and strengthening the ongoing tree breeding and improvement activities. The application of plant tissue culture in forestry facilitates rapid exploitation of genetic vigor and heterosis through mass multiplication and ensures the availability of improved and quality planting materials. Most plantation programme in the country suffers due to lack of quality planting material which results in poor and uncertain productivity. For this purpose, the tissue culture technology would play a vital role in mass multiplication of genetically improved and quality improved planting materials through true to type clonal plants. This besides, the tissue culture technologies also extend the scope of development of disease free plants which is one of the major demands for most single species based plantation forestry programme. Several tree species have been accredited with wide range of secondary metabolites which may have profound utility in pharmaceutical sector and the tissue culture technology facilitate in-house production of secondary metabolites. Above all, tissue culture technology extends a greater scope of *ex-situ* conservation of Rare, Endangered and Threatened, (RET) Species through *in-vitro* approaches. The information provided in this chapter will act as a master piece of the information for the readers not only for the academic purposes but also to practice commercial ventures through tissue culture technology deploying commercially viable and acceptable tree species. This technology ensures continuous production of plantlets throughout the year and may help to achieve the objectives envisaged by various stakeholders towards mass multiplication and conservation of elite forest genetic resources.

15

FOREST BIOTECHNOLOGY

INTRODUCTION

Biotechnology is an integration of biological and technological components. During the last three decades, the subject of biotechnology has gained significant importance in the field of biological science towards development of process and products. The biotechnology has been very well deployed and utilized in medical sciences towards addressing various issues in human health system. Then the technology has its wider application in industries using living organism as a source of developing products of industrial applications. This has resulted in intensive adoption of biotechnology thus resulted in development of wide range of products which are antibacterial, antifungal and antiviral utilities.

The techniques of biotechnology provides the possibility of wide range of products and compounds and found greater utility in chemical, food, pharmaceutical, environmental protection, agricultural productivity and pollution abatement. This significant application in wide range of biological sciences has received attraction towards agriculture thereby the science of agricultural biotechnology has received greater attention. The biotechnology application in agriculture has been found enormous utility in three different domains like tissue culture, plant molecular biology and plant biochemistry. Protocols were developed for mass multiplication of wide range of agriculture and horticulture crops. Similarly the plant molecular biology has been deployed to characterize the diversity pattern, marker assisted selection and DNA sequencing of various crop varieties which have helped the breeders towards rapidly adopting their breeding strategies.

Due to successful developments in agriculture and horticulture sectors, the tree breeders in the country have also adopted various biotechnological tools towards improvement and conservation of wide range of forest genetic resources. The major tools and techniques like tissue culture for mass multiplication, molecular biology tools like DNA markers for diversity and finger printing studies, genetic engineering for development of pest and disease tolerance, herbicide tolerance, varieties for bioremediation, modified trees for wood quality etc are predominantly used in forestry sector. Similarly the biotechnology mediated chemical and biochemical characterization leads to inventory and development of wide range of products of pharmaceutical importance. This chapter elaborates all major applications of biotechnological tools in forest trees.

DEFINITION OF TERMINOLOGIES

Biotechnology : Biotechnology is the broad area of biology involving living systems and organisms to develop or make products. (*or*)

Any technological application that uses biological systems, living organisms, or derivatives thereof, to make or modify products or processes for specific use.

Genetic engineering : Genetic engineering, the artificial manipulation, modification, and recombination of DNA or other nucleic acid molecules in order to modify an organism or population of organisms.

Secondary metabolites : Secondary metabolites are organic compounds produced by bacteria, fungi, or plants which are not directly involved in the normal growth, development, or reproduction of the organism.

DNA : DNA is Deoxyribo Nucleic Acid, a self-replicating material which is present in nearly all living organisms as the main constituent of chromosomes. It is the carrier of genetic information.

Markers : A character or a gene which, by virtue of linkage, can be used to indicate the presence of another gene.

or

A marker is a gene or DNA sequence with a known location on a chromosome that can be used to identify individuals or species.

PCR : PCR - Polymerase Chain Reaction

PCR is a method of making multiple copies of a DNA sequence, involving repeated reactions with a polymerase.

or

Amplification of DNA through cycles of denaturation (by heating) and synthesis of new strands.

Plasmid : A plasmid is a small, circular piece of DNA that is different than the chromosomal DNA, which is all the genetic material found in an organism's chromosomes. It replicates independently of chromosomal DNA.

Vector : Vector is a DNA molecule used as a vehicle to artificially carry foreign genetic material into another cell, where it can be replicated and/or expressed. E.g., plasmid, cosmid, Lambda phages.

A vector containing foreign DNA is termed recombinant DNA.

Enzymes : An enzyme is a substance produced by a living organism which acts as a catalyst to bring about a specific biochemical reaction.

or

An enzyme is a substance that acts as a catalyst in living organisms, regulating the rate at which chemical reactions proceed without itself being altered in the process.

Genome : The complete genetic complement of an individual organism.

Gene : A segment of DNA giving rise to a particular enzymic product and having a particular effect on phenotype.

Microsatellite : DNA sequence composed of a tandem replication of a simple short sequence, occurring in the genome of many higher organisms.

Nucleotide : A compound comprising a nucleoside linked to phosphoric acid. When polymerized give rise to nucleic acids.

Primer : A short oligonucleotide that pairs to a region of single stranded template oligonucleotide.

Transformation : The incorporation of new DNA into a cell's genome.

SIGNIFICANCE AND ROLE OF BIOTECHNOLOGY IN FORESTRY SECTOR

The application of biotechnological principles in forestry sector is termed as forest biotechnology. Though biotechnology has a very vast application in wide range of sciences and its application in forestry sector is hardly one to two decades old. Again the biotechnological application in forestry sector compared to other sciences is very limited. The major areas of biotechnological application in forestry sector are furnished below

a. Tissue culture and micro propagation

Several forestry and commercially important agro forestry tree species have been mass multiplied through plant tissue culture based biotechnological approaches. This has addressed the issues of rapid multiplication, availability of quality planting materials and maintenance of clonal fidelity. Successful protocols are now available for species like Teak, Sandal, Casuarina, Eucalyptus, Melia, Acacia, Albizia, Bamboo and other medicinal plants. The details of tissue culture along with advantages and limitations are furnished in chapter 14.

b. Molecular markers and its application

Application of molecular markers in forestry sector has witnessed and found utility in terms of the following areas

- Quantification of genetic diversity and similarity between the populations.
- Finger printing of forest genetic resources.
- Identification of sexes, mating system and gene flow in populations.
- Maintenances of germplasm and clonal fidelity.

- Identification of traits of importance and incorporation in genetic engineering.
- Marker assisted selection.
- Genetic mapping.
- Characterization of genetic resources.
- Diversity analysis and biodiversity conservation

c. Development of genetically modified trees

Development of genetically modified trees for various needs is one of the potential areas of biotechnology which involved manipulation of DNA or genetic material inside the cells of living organisms. This genetic engineering has witnessed wider application in agriculture leading to productivity improvement and resistance towards biotic and abiotic factors. During the last four decades of research, wide range of GMO's are now currently available. This genetic engineering technology has also been adopted by wide range of tree breeders and forest biotechnologist which resulted in genetic manipulation in tree species. Modification of wood quality through lignin and cellulose alteration, manipulation of density through gene changes, development of biotic and abiotic resistance varieties in poplars, Eucalyptus, pines, etc., are the major evidences of genetic engineering application in forestry sector. Unlike agriculture most GMOs in tree crops are developed through *Agrobacterium tumefaciens* mediated gene transfer method. Similarly *Agrobacterium rhizogenes* has been used to augment the rooting efficiency of several species through gene transfer mechanism.

d. Biotechnology Mediated Ecological Restoration

Biotechnologies mediated genetic engineering technology leads to modification of several species which are amenable for ecological restoration. The very classical example is the modified species of American chestnut, English Almond, southern pines etc., which have been utilized to restore the pest damaged landscapes.

e. Dendro Remediation

Biotechnology mediated gene transfer mechanism has been deployed in several tree species towards remediation of heavy metal and other organic pollutant affected area. The genetically modified trees play a vital role in remediation and reclamation of heavy metals and other organic pollutants. This concept of reclamation through genetically modified trees is pronounced as dendroremediation.

Example: The transformed yellow poplar (*Liriodendron tulipifera*) expresses bacterial mercuric reductase and grows vigorously in toxic levels of ionic mercury. This enzyme converts the highly toxic ionic mercury to much less toxic elemental form and has received significant attention towards deploying genetically modified poplar for dedroremediation.

f. Microbial Remediation

The ability of several microorganisms to accumulate heavy metals and their potential use in the decontamination of environment is very well witnessed. The heavy metals are very dangerous which increases due to anthropogenic pressure and disrupts normal biogeochemical thus leads to environmental degradation. The industries like chrome tannining, electroplating, dyes, paints, preservatives, textile etc are the major sources of heavy metal pollutants. The traditional methods of heavy metals removal are ineffective and the biotechnology mediated microorganisms through biological removal process has created significant attraction and is deployed in several heavy metal contaminated areas.

g. Secondary Plant Metabolites

The plant cells produce two kinds of metabolites viz., primary metabolites and secondary metabolites. The primary metabolites involve directly in growth and metabolism like proteins, carbohydrates and lipids. The secondary metabolites are mostly of the end products of primary metabolites like alkaloids, phenols, steroids, tannins, lignins, essential oils etc., and they are not involved in metabolic activity. These secondary metabolites are mostly accumulated by plant cells in smaller quantity and biotechnology helps in amplification of these products for industrial utility. Trees in general and some herbs and trees associated with forest ecosystem in particular are acting as a wide range of resources for these metabolites and these species received potential attraction towards development of wide range of products through biotechnology mediated cell suspension cultures.

Secondary metabolites like Taxol, Quinine, Arjunalic acid, Jatrophin camptothecin, santalin, santalol, azadiractin etc., identified and isolated from wide range of tree species witnessed the commercial application of biotechnological tools for secondary metabolites production. Many more tree species accredited with wide range of metabolites need to be identified and commercialized for which biotechnological approaches have wider scope and significance.

h. Biofuels

Biotechnology play a vital role in developing technologies for producing biofuel to cater to the needs of growing energy demands coupled with reducing the impacts of environment due to use of fossil fuels. The conversion of wood based lignocelluloses biomass into a ethanol is one of the major strategies to blend at least 20% into ethanol. Several species are identified as a potential raw material for bioethanol production

Example: *Melia dubia* (MTP 2) A HYSR Clone is identified as a potential species for bioethanol production.

Similarly biodiesel is extracted from wide range of tree borne oil seeds like Jatropha, Karanj, Madhuca, Callophyllum etc., extend greater scope of biotechnology mediated biodiesel production. The clonal technology in the production side and the transesterification process in the prossessing side are the major biotechnology interventions towards biodiesel production.

MAJOR BIOTECHNOLOGICAL TOOLS DEPLOYED IN FORESTRY SECTOR

Though biotechnology has wider applications in forestry but has found extensive utility only in the following four sectors as indicated in the Fig. 15.1 which are discussed in detail.

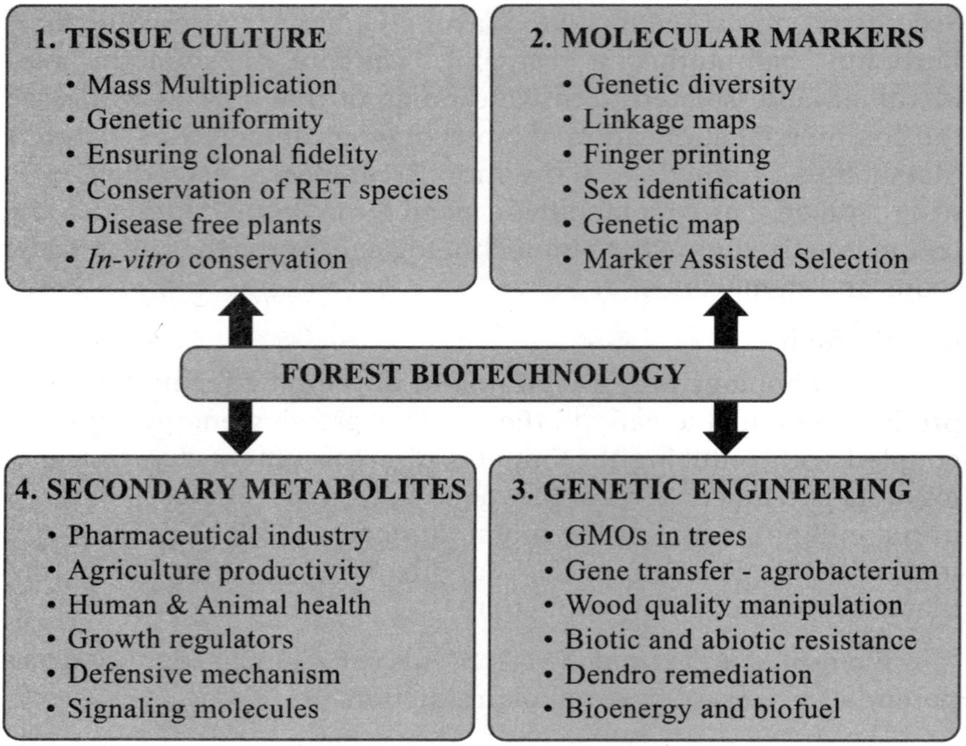

Fig 15.1. Biotechnological Tools Applied in Forestry Sector

1. TISSUE CULTURE

The techniques of tissue culture have found increased utilization in several tree crops and successful protocols are now available for wide range of tree crops of timber, pulp and paper, plywood, biofuel and bioenergy utility. The process of direct and indirect organogenesis has been deployed successfully in several tree species. The basic concepts, methods and various stages of plant tissue culture deployed in major tree species are elaborately discussed in chapter 14 of this book.

2. MARKERS

Markers particularly the DNA based molecular markers are important tool for tree breeding and improvement. The tree breeding and improvement programme are traditionally a long term process due to the larger size of tree structure coupled with long rotation of tree species. In this context, the markers particularly the molecular markers have played a vital role in advancing tree breeding approaches and reduce the rotation period of tree species coupled with resistance towards various biotic and abiotic factors. Hence it is very essential to learn about various types and kinds of markers.

Markers are those characters which can be easily identified or recognised. The markers related to morphological variations are called phenotypic markers and the markers related to variations in DNA fragments generated by restriction endonuclease enzymes are called as genetic or DNA markers.

Types of markers

There are four types of markers which are used in tree breeding and improvement programme. They are

i. Morphological or Phenotypic Markers

ii. Biochemical or Isozyme Markers

iii. Cytological or Chromosomal Markers

iv. DNA or Genetic Markers

i. Morphological or Phenotypic Markers

Traditionally the tree breeders use Morphological or Phenotypic Markers in improvement of most of the tropical species. These markers are characters which are related to variation in size, shape, colour and surface of various tree parts. These markers generally refer to the available genes which have impact on the morphology of tree species. The genes that affect tree form, structure, resistance etc., have been analysed in several tree species.

E.g.: Selection of timber species with the characters of straight cylindrical bole.

Selection of TBO's having large crown with more number of branches.

Presence of lenti cells on the bark of Casuarina.

Presence of crocodile bark in Red sanders.

Presence of smooth and rough bark in *Melia dubia.*

ii. Biochemical or Isozyme Markers

These are markers related to variation in proteins and

aminoacids banding pattern. In this marker, a gene encoding a protein is extracted and observed like Isozyme analysis.

E.g.: Isozyme analysis in Teak genetic resources.

iii. **Cytological or Chromosomal Markers**

These markers are related to variation in chromosome number, shape, size and banding pattern. This marker refers to a chromosomal banding produced by different stains.

E.g.: Gene banding.

iv. **DNA or Genetic Markers**

It is otherwise called as molecular markers wherein a gene or other fragment of DNA whose location in the genome is known. It refers to any unique DNA sequence which can be used in DNA hybridization, PCR or restriction mapping analysis to identify the sequence. Among all the four types of markers, the application of DNA markers is gaining significant attraction among tree breeders and forest biotechnologists because of the following advantages

- DNA markers are highly polymorphic.
- Most of the markers are co-dominant markers because of simple inheritance.
- These markers occur profusely throughout the genome.
- DNA markers are not influenced by environmental effects.
- Molecular markers exhibit a minimum Pleiotropic effect.
- Above all, the molecular markers are easy and fast to identify and detect.

Different types of Molecular Markers

The markers are classified into two types based on PCR and Non-PCR methods.

Marker based on Non-PCR method

i. **RFLP- Restriction Fragment Length Polymorphism**

It is a co-dominant marker and it is a first DNA based genetic marker developed. This marker can detect the coupling phase of DNA molecules because DNA fragments for all homologous chromosomes are detected. They are very reliable in linkage analysis and helps the breeder to determine whether the linked trait is present in a homozygous or heterozygous stage in the individual. RFLP markers are very powerful tool for comparative and synteny mapping. (Xu 2010)

ii. PCR based DNA markers

The PCR based marker has been recognised as one of the most important biological discoveries invented during 20th century by Kerry Mullis which gave him a Nobel prize (Mullis 1990). This PCR enables production of large amount of a specific DNA sequence without the cloning process, starting with just a few molecules of the target sequence. The PCR reaction has involved three basic steps:

- Denaturation of double stranded DNA template.
- Annealing of a pair of primers to the region of amplification.
- Amplification deploying Taq polymerase enzyme.

  The completed sequence of three step is called a cycle and at the end of first cycle two new double stranded molecules are produced from original template. This newly produced molecules serve as a template for subsequent cycles which lead to geometric amplification of molecules. The process of PCR is carried out through a specially designed instrument called thermocycler or PCR machine.

Major PCR Based Markers

The most widely used PCR based markers are discussed below.

a. Random Amplification of Polymorphic DNA (RAPD)

RAPD markers are one of the most commonly used markers in tree breeding due to its simplicity and less expensive. These markers produce DNA profiles of varied complexity depending on the primer and the template. Another important advantage of this marker is no clear DNA sequences information is needed. The RAPD markers are very well exploited in tree breeding of various tree species as indicated below:

S.No	Species	Primers used	Reference
1.	Teak	OPO-2, OPO-9, OPO-13, OPO-20, OPR-2, OPR-6, OPR-10, OPR-11, OPR-15, OPM-11, OPM-13, OPAP-9, OPAP-7, OPAP-13, OPAW-16, OPY-15 & OPY-17	Parthiban, 2001
2.	Simarouba	OPAB 05, OPAB 09, OPAB 15, OPAK 05, OPAK 06, OPAM 16, OPAM 19, OPR 02, OPR 10 & OPR 11	Sekar, 2003
3.	Eucalyptus	OPT 1, OPT 4, OPT 8, OPT 15 & OPF1	Vennila, 2009
4.	Jatropha	OPF 1, OPF 6, OPF 7, OPF8, OPF10, OPF 11, OPF 12, OPF 14, OPX 9, OPX 20, AP 12, OPC 1. PQS 4, OPE 19, R 3, OPAA 1, OPAA 5, OPAA 6, OPAA 7, OPAA 12, OPAA 14, OPAA 16 & OPAA 17	Dhayananthan, 2011
5.	Sissoo	OPA-1, OPG-11, OPB-15, OPBB-13 & OPBB-19	Dhixya Deve, 2014

b. Amplified Fragment Length Polymorphism (AFLP)

These are advanced markers but poorly exploited in tree improvement programme. AFLP markers has a greater applications in diagnostics, pedigree analysis, forensics, parental heritage and also in finger printing (Pereira et.al., 2010). The markers have been exploited in following species like

- Larch (Arcade *et al.,* 2000)
- Neem (Singh *et al.,* 2002)
- AFLP marker was used to genetically map a disease resistant gene in Poplar (Cervera *et al.,* 1996)
- Genetic linkage maps using AFLP's have also been constructed in different Eucalyptus species (Marques *et al.,* 1998).

c. Simple Sequence Repeat (SSR)

These markers are called as micro satellites or simple sequence repeat markers were first developed to use in genetic mapping in humans (Litt and Luty 1989). These markers are short segments of DNA that are derived from short (less than 6 base pairs) tandemly repeated sequences like (GA) N, (AAT) N and (GT) N. The micro satellite markers are used more robustly because of their co-dominant expression, multi allelism and PIC value. Micro satellite markers have been used and characterized in several species as indicated below

S.No	Species	Primers used	Reference
SSR (Simple Sequence Repeats)			
1.	Redsander	CCB 1, CCB10, CCB7, PGM10, PGM102, PGM106, PGM109, PGM16, PGM3, PGM45, PGM5, PGM82, PKS18, PKS26, PKS30, CcM0257, CcM0948, CcM0008, CcM0039 & CcM1026	Karthick, 2019
2.	Eucalyptus	-	Brondania *et al.*, 1988
3.	Rubber	-	Fing *et al.*, 2009
ISSR (Inter Simple Sequence Repeats)			
1.	Sandal	UBC 807, UBC 808, UBC 811, UBC 814, UBC 824, UBC 834, UBC 841, UBC 888 & UBC 890	Krishnakumar, 2017
2.	Kadam	UBC 807, UBC 808, UBC 809, UBC 811, UBC 814, UBC 823, UBC 824, UBC 826, UBC 834, UBC 841, UBC 842, UBC 854, UBC 855, UBC 856, UBC 857, UBC 860, UBC 861, UBC 864, UBC 866, UBC 867, UBC 880, UBC 885, UBC 886, UBC 887, UBC 888, UBC 890 & UBC 891	Thiruniraiselvan, 2018

S.No	Species	Primers used	Reference
3.	Pungam	UBC808, UBC809, UBC810, UBC811, UBC823, UBC834, UBC835, UBC844, UBC848, UBC881 & UBC890	Palanikumaran, 2015
4.	Callophyllum	UBC808, UBC809, UBC810, UBC811, UBC823, UBC834, UBC835, UBC844, UBC848, UBC881 & UBC890	Palanikumaran, 2015
5.	Jatropha	I1, I2, I3, I4, I5, I6, I7, I8, I9 & I10	Senthilkumar *et al.*, 2009

d. Variable Number Tandem Repeat (VNTR)

These markers are also called as mini satellite and are used for various purposes like DNA finger printing, diversity analysis and linkage map construction. The marker in this type is a section of DNA that consists of a short series of bases (10 - 60 base pairs) in length. These markers contain repetitive, generally rich variant types. These types of variant repeats are tandemly intermingled which makes this marker ideal for studying DNA turnover mechanism and found extensive utility in genetics.

e. Cleaved Amplified Polymorphic Sequence (CAPS)

This marker and the associated technology expresses similarity to RFLP method and known as (RFLP) PCR technique. In this marker, the DNA fragments are amplified using specific primers digested by restriction enzymes. The sequence polymorphism results due to cutting of products in different places and these variants are released as length differences in the system. This marker found utility in organism specific substance.

f. Single Nucleotide Polymorphisms (SNPs)

These markers are known as third generation markers found extensive usage in genomic studies for individual genotype. These markers are DNA sequence variation that occur due to point mutation when a single nucleotype (ARTRCRG) in the genome sequence is altered. Successful application of SNP genotyping was witnessed in Pinus (Neale, 2007).

g. Expressed Sequence Tag (EST)

EST marker play a vital role to identify gene transcripts, gene discovery and gene sequence. However the application of this marker in forestry is dismally modest.

3. GENETIC ENGINEERING

Genetic engineering is a gene transfer method and witnessed wider application in agriculture sector but its application in forestry is limited to only few tree species of temperate and tropical origin. Most of the GMOs in tree crops are facilitated through *Agrobacterium tumefaciens* mediated gene transfer method.

Applications of Genetic Engineering in Forestry

The genetic engineering has wider applications in forestry in the following sector:

i. Improvement of wood quality particularly lignin and density modification. This will improve the wood quality both interms of physical and chemical properties and help to cater the needs of various wood based industries.

ii. Genetic engineering helps to improve the industrial process through lignin modification which lead to reduction in energy, water consumption and reduced impact on pollutants.

iii. Genetic engineering helps to develop pest and disease resistance in tree crops. Successful results have already been reported in following species

- Resistance to the fungal pathogen *Septoria musiva* in poplar expressing chitin-binding proteins Ac-AMP1.2 and ESF12 (Liang *et al.* 2002)
- Resistance to plum pox virus in *Prunus domestica* following expression of the viral coat protein (Ravelonandro *et al.* 1999)
- Resistance to the cottonwood leaf beetle (*Chrysomela scripta*), among poplars following their genetic modification with Bt toxin genes (James *et al.* 1999)

iv. Genetic engineering has found wider application towards development of herbicide tolerant poplar which lead to increased eco-friendly products. The transgenic poplars with higher levels of Glutamylcysteine and Glutathione (GSH) levels compared to the wild poplars exhibited higher tolerance towards herbicide.

v. The genetic engineering trees are also amenable for wide range of edapho climatic conditions and hence they play a vital role in ecological restoration.

vi. Genetically modified trees are also known to produce wide range of products like hormones and drugs which have profound utility in pharmaceutical industry. One of the classical example is the engineered rubber trees in Malaysia is able to secrete human serum albumin. (Langridge, 2000)

vii. One of the major application of genetically modified trees is yield improvement. Significant productivity improvement has been witnessed in various poplar species compared to the wild species and engineered species.

viii. The application of GM has made very significant impact in facilitating and improving the conventional breeding methods in terms of early flowering, rapid maturity, profused seed formation, higher yield etc.,.

ix. The genetic engineering also helps in augmenting the rootability of several tree species through *Agrobacterium rhizogens* based gene transfer mechanism and strengthen the activities of clone based plantation development programme both in the tropical and temperate region.

4. SECONDARY METABOLITES FROM FORESTS

Secondary metabolites are natural metabolic products with the heterogeneous group of compounds which are not involved in metabolic process leading to growth of producing organism but have several functional roles like defencing mechanism, Signalling molecules, symbiosis, transport, competition etc., These metabolites present in small quantity in a plant system are exploited by human beings to cater to the needs of human health, nutrition, improving agriculture productivity and growth promoters of plants and animals. Due to these significant applications

Secondary metabolites obtained from plants through biotechnological approaches have become great significance. Among the plant based products, the forest trees and their associated annuals and shrubs are accredited with wide range of secondary metabolites and hence application of biotechnological tools in secondary metabolites production is of great value.

Different types of metabolites

It is reported that over 21,40,000 secondary metabolites are known at present and these secondary metabolites are classified into 5 major groups (Durairaj *et al.,* 2018)

i. **Terpenoides and steroids**

There are over 35000 compounds of terpenoides are identified which are derived biosynthetically from isopentyl diphosphate. Terpenoides expressed different varieties of undulated structures and the steroids have a common tetrocyclic carbon Skelton.

ii. **Alkaloids**

These metabolites consist predominantly of basic amine groups and derived biosynthetically from amino acids. It is estimated that over 12000 compounds of alkaloids are known as of now.

iii. **Fatty acid derived substances and polyketides**

These metabolites are biosynthezed from simple acyl precursors like propionyl coA, Asytl CoA and Methylmalonyl CoA. It is estimated that over 10000 compounds of this category of metabolodes are known at present.

iv. **Non ribosomal polypeptides**

These group of metabolites are predominantly amino acid derived compounds and synthesized biologically through a multifaceted enzyme complex without direct RNA transcript.

v. **Enzyme cofactors**

These are low molecule enzyme compounds and are non protein in nature.

Major Tree species as source of metabolites

Wide range of tree species and their associated herbs and shrubs are known to contain different metabolite products which are synthesized through conventional method or cell immobilization and also through cell suspension and hairy tip cultures. The metabolites present in multiple tissues of various tree species can be synthesized through callus mediated cell suspension cultures. The major metabolites identified from various tree species through conventional and also through biotechnological mediated process are furnished below

S. No.	Species	Secondary metabolites
1	Cinchona	Quinine and Quinidine
2	*Coffea arabica*	Purine alkaloids
3	*Panax ginseng*	Ginseng
4	Agave	Sisalana And Hecogenin
5	*Catharanthus roseus*	Vinblastin and vincristine
6	Rauwolfia	Reserpine
7	Jatropha	Curcine
8	*Digitalis lanata*	Digoxin, digitoxin
9	*Taxus buccata* and *T.brewifolia*	Taxol
10	*Pterocarpus santalinus*	Santalin
11	*Terminalia arjuna*	Arjunalic acid
12	*Gloriosa superba*	Colchichine
13	*Calophyllum inophyllum*	Amantoflavone
14	*Mappia foetida*	Camptothecin
15	*Terminalia bellerica*	Lignine
16	*Anogeissus latifolia*	Anolignin
17	Gymnocladus	Triterpenoid
18	*Azadirachta indica*	Nimbin, nimbidin and other 44 alkaloids
19	*Croco sativus*	Crocin
20	*Bixa orellana*	Bixin
21	*Rubia tinctorum*	Alizarin and purpurine
22	*Achras sapata*	Chicle
23	*Aquilaria malaccensis*	Oleoresin
24	*Santalum album*	Santalol

Functions and roles of Secondary metabolites

Secondary metabolites play a vital role in the following areas.

- Major Secondary metabolites extend a defensive role against the wide range of biological organisms there by help to protect the host plant.
- Metabolites also play a role as a transporting agent for metals.
- Some of the Secondary metabolites are acting as an agent for symbiotic relation with other organism and helps in completion of the symbiotic process.
- Some Secondary metabolites are acting as the reproductive agents for several biological organisms.
- They also play as a source of differentiation effectors
- Some metabolites extends signalling mechanism which helps to exchange the communication between the organism which lead to the major activities of breeding, pollination, defencing, attraction etc.
- The major role of Secondary metabolites is towards application in human and animal health. Several Secondary metabolites exhibits antibacterial, anti fungal and anti viral properties and found major utility in pharmaceutical industry.
- Metabolite identified from trees also acting as a source of anti-tumor characteristics and helps in curing the cancerous diseases.
- Some of the Secondary metabolites are an excellent source of products for animal health.
- Several metabolites are acting as a source of bio pesticides and bio herbicides and helps in agriculture productivity.
- Some of the metabolites are also acting as an excellent growth regulator.

SUMMARY

Application and exploitation of biotechnological tools are rapidly increasing across different sectors like agriculture, microbiology, industries, pharmaceuticals, engineering etc,. However, its application in forestry sector particularly in tropical forest is only limited to tissue culture, molecular biology, genetic engineering and products of secondary metabolites. Forests are excellent resource materials for various wood based industries till the recent past; however due to policy and legal implications, encouraged deployment of plantation forestry programme which demands application of biotechnological tools. The development of tissue culture protocol for wide range of tree species has strengthened the clonal plantation development programme of various wood based industries which leads to enhanced productivity coupled with reduction in harvesting period. The genetic manipulation through gene transfer mechanism has developed successful GMO tree in poplars, pines and eucalyptus which cater to the needs of wood quality improvement, resistance towards pest and diseases and tolerance to herbicidal application. In the recent past, integration of biotechnological tools with forestry has led to the development of wide range of plant products like alkaloids, steroids, terpenoides etc which found greater utility in pharmaceutical industry, food industry and agriculture sector. The forest biotechnology has also strengthened the traditional breeding activities and helped the breeder to achieve the breeding goals in a short time and speedy process. The genetic diversity analysis, finger printing and Marker assisted selection have helped the breeder to overcome the obstacles of long term traditional breeding process. In a holistic view, the biotechnological tools are finding excellent utility in forestry sector, however it needs intensification and more adoption of the biotechnological tools to improve the productivity, augment the wood quality and conserve the forest genetic resources.

16

TREE VARIETAL DEVELOPMENT PROCESS

INTRODUCTION

India is one of the robust consumers of wood and wood products but there is no concomitant organized plantation development programme. It is estimated that the country would need over 152 million m^3 of wood in the form of short rotation and long rotation species to cater to the raw material needs of over 18 different organized sectors of wood based industries. In addition to this, another 300 million m^3 of wood is needed to meet the domestic wood demand coupled with the raw material needs of unorganized sectors of wood based industries. The existing natural forest is almost closed for felling due to legal and policy regulation coupled with complete ban of felling in the natural forest by the Honorable Supreme Court. The growing wood demand coupled with diminishing supply from natural forest have attracted intensive promotion of organized plantations and agroforestry systems. The National Forest Policy (1988) has directed all wood based industries in the country to develop their own plantations by establishing suitable linkages with farmers. These developments have made wood based industries to deploy tree breeding methods to develop high yielding varieties amenable both for agroforestry and for multifarious industrial utility. Several research institutes, organizations and wood based industries are involved in tree breeding and improvement programme in over 20 different species of commercial importance across the country and are at various stages of development. Tamil Nadu Agricultural University is one of the pioneering institutions in the country involved in tree breeding programme in the year 1976 onwards and brought out improved tree varieties in several species like Acacia, Simarouba, Casuarina, Eucalyptus, Melia, Jatropha and Kadam. Several improved HYSR clones in other species like Teak, Toon, Mahogany, Acrocarpus, Albizia, Leuceana etc., are in the pipeline. The current chapter has been designed to disseminate the tree breeding methods followed for evolving a new tree variety which may be of great use to the tree breeders of the country and other stakeholders involved in tree breeding programme.

MAJOR TREE VARIETIES RELEASED BY TNAU

S.No	Species	Variety released through SVRC	Year of release
1.	*Acacia holosericea*	Mankathu velamaram	1989
2.	*Simarouba glauca*	Simarouba MTP 1	2000
3.	*Casuarina equisetifolia*	Casuarina MTP CA 1	2008
4.	Casuarina Hybrid	TNAU Casuarina MTP 2	2011
5.	*Melia dubia*	Malai Vembu MTP 1	2012
6.	*Eucalyptus camaludensis*	Eucalyptus MTP 1	2013
7.	*Melia dubia*	Malai Vembu MTP 2	2018
8.	*Neolamarckia cadamba*	Kadam MTP 1	2019

SALIENT FEATURES OF THE TREE VARIETIES RELEASED BY FCRI, TNAU

S.No	Varieties	Salient features
1.	*Acacia holosericea* Mankathu velamaram	• It is an introduction from Africa • Fast growing tree • Suitable for firewood and biomass power generation • High calorific value of over 4000 K cal / kg • Short duration variety amenable for harvest in 3 years • High yielder with a production potential of 100 tons / ha • Suitable for waste land Afforestation

S.No	Varieties	Salient features
2.	Simarouba MTP 1	• This variety is evolve through AICRP project on underutilized crop funded by ICAR • It's a selection through series of seed source evaluation • It is amenable for cultivation across different land use system • The tree yields an oil which is edible due to acceptable oil quality • The timber is useful as raw material for match splints production.
3.	Casuarina MTP CA 1	• First high yielding short rotation clone in the species • Yield potential is 100 tonnes per ha. • Duration - 36 months • Pulp yield over 48 %
4.	TNAU Casuarina MTP 2	• Hybrid clone • High yielder with a yield potential of 150 tonnes per ha • Amenable for pulp and pole industries • Short duration variety with a rotation ranging between 30-36 months.
5.	Malai Vembu MTP 1	• It is an alternate plywood species. • Amenable for harvest in 5-6 years • Veneer recovery is over 58 %. • Increased veneer quality • Yield – 150 tonnes / ha
6.	Eucalyptus MTP 1	• It is a improved clone with a pulp yield of over 48 % • Amenable for dry land agroforestry • Kappa Number – 18-20 • Yield - 130 tonnes per ha • Rotation – 5 years

S.No	Varieties	Salient features
7.	Malai Vembu MTP 2	• Clonal variety • Suitable both for pulp and ply wood production • For pulp wood, the clone can be harvested at 24 months with an yield potential of 100 tonnes per ha • Pulp yield is over 50 % and hence identified as one of the superior clones for pulp and paper production. • For plywood the clone can be harvested between 4 and 6 years • The yield potential for plywood is 150-200 tonnes / ha
8.	Kadam MTP 1	• Clonal variety • The pulp yield of this variety is more than 44 % and the kappa number is above 20 % • The veneer recovery is more than 52% • The leaves are excellent fodder - Protein (16 – 22 %) • Productivity o 135 tonnes / ha in 3 years for pulp wood o 150 – 175 tonnes / ha in 5 years for plywood

TREE VARIETY DEVELOPMENTAL PROCESS

As indicated earlier, FC & RI, TNAU has released 8 different tree varieties and another 12 tree varieties representing 8 species are in various stages of evaluation. During the last three years, the institute has released two major varieties one each in Melia and Kadam which received significant attraction from all levels of stakeholders. The major tree variety developmental process followed for these two varieties are furnished as a reference method for the tree breeders to understand and to learn the tree varietal developmental process.

i. VARIETAL DEVELOPMENTAL PROCESS FOLLOWED FOR *Melia dubia* MTP 2

Parentage	:	**Selection from Existing population**
Breeding method	:	Clonal selection
Breeding Objectives	:	To identify High Yielding and Short Rotation (HYSR) Clones amenable for pulp and plywood production
Breeding Approach	:	1. Preliminary - Seed Source Evaluation Trial - 1st Generation 2. Clonal Test - 2nd generation 3. Amplified clonal test - Farmer / Industry / Institution - 3rd generation 4. Industrial quality test for wood quality

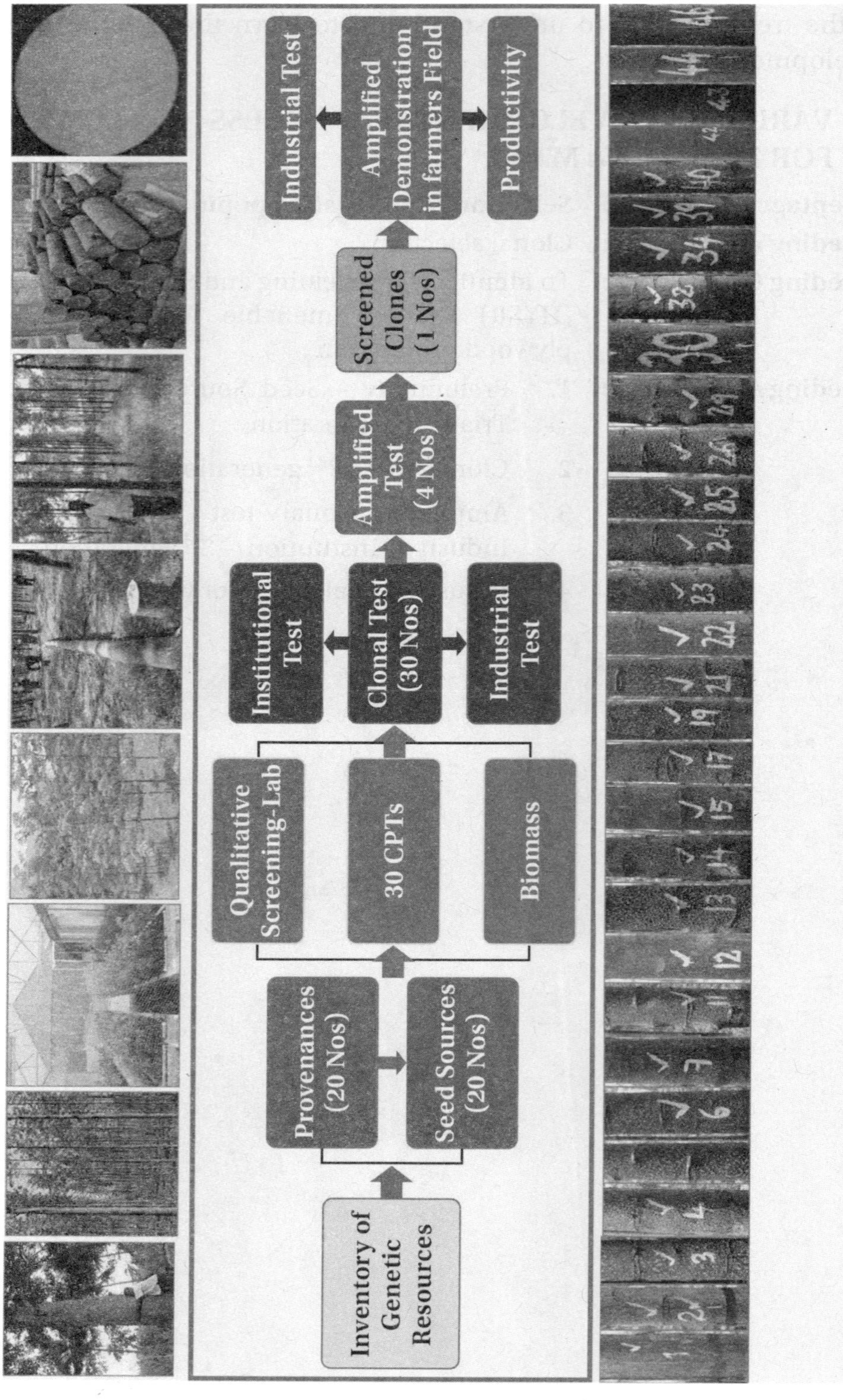

Fig. 16.1. Breeding Plan and Approach

PROCESS OF GENETIC TESTS AND RECOMMENDATIONS

Regular clonal test	:	• Thirty plus trees were identified from a seed sources and provenance tests based on growth evaluation after 5 years of establishment. • The selected trees were harvested and assessed for total biomass and wood quality. • These trees were cloned using apical shoots obtained from coppices and are termed as clones. • These clones were evaluated in three locations and the superiority of clone MD 26 has been evidenced.
Amplified clonal Test (MLT)	:	The clone MD 26 has been amplified in nine farmer's fields, two industrial fields and one at FCRI campus at an area of 1 acre per farm and assessed for growth attributes for the period from 1 to 4 years to fix the rotation and the yield.
Amplified clonal Test with intercrops	:	The same clone has been evaluated under agroforestry system deploying pulses and vegetables to assess the compatibility, yield and income.
Demonstration cum evaluation test (FLD)	:	Final demonstration in thirty farmers field at an area of 30 acres to validate the potential of the proposed clone

FINAL CLONE SCREENED FOR ADOPTION

A Clonal Selection from *Melia dubia* viz., MD 26 has been screened through systematic clonal selection followed by evaluation, amplified clonal test and demonstration for a period of ten years. Based on the systematic clonal evaluation, the variety MD 26 has been identified as a fast growing multipurpose tree species amenable for varied industrial utility and different agroforestry systems. The salient features of the variety are furnished below.

a) Pulp qualities

- Pulp yield : 48–49.6 %

- Basic Density : 400 - 530kg m^{-3}
- Hollo Cellulose : 70 – 75 %
- Kappa number : 16 - 22 %
- Burst index : 3.5 – 4.5 K Pa m^2 g^{-1}
- Tensile strength : 70 – 95 Nm g^{-1}
- Tear index : 7.0 – 8.5 m Nm^2 g^{-1}

b) Plywood Properties

- Bark Thickness : 4.42 mm
- Veneer yield : 62.56 %
- Veneer quality : Face quality
- Shrinkage (%) : 5.24
- Bonding : Excellent
- Mat thickness (mm) : 11.94

c) Thermo chemical properties

- Calorific value : 3820 cal/kg
- Ethanol yield (UT) : 0.37± .0 (g/g)
- Ethanol yield (T) : 0.47± .0 (g/g)

d) Leaf Quality

- Moisture Content : 62.41 %
- Dry Matter production : 37.55 %
- Ash Content : 7.40 %
- Crude fat : 2.96 %
- Crude Protein : 16.95 %
- Carbohydrate : 16.00 %

PROCESS OF VARIETY RELEASE

The identified superior clonal variety (MD 26) has been presented at the Annual Research Meet and Scientific Worker Conference held during the years 2015 & 2016. Based on the presentation, the Annual Research Meet – Forestry held during the year 2015 & 2016 recommended the proposal for approval in the Scientific Worker Conference. The proposal was approved in the Scientific Worker Conference held during 2015 & 2016.

Based on the recommendation of ARW & SWC, a superior clone MD26 has been placed before the University Varieties and Technologies release screening committee (UVTRSC) for approval on 11.12.2017. The salient features of the variety along with various breeding methods adopted have been presented before the committee and the committee has approved and recommended for submission to State Variety Release Committee (SVRC).

The variety is then placed before the state variety release committee held on 28.12.2017 chaired by the Agricultural Production Commissioner and Principal Secretary to Government of Tamil Nadu, Agricultural Department in the presence of SVRC Members. The tree breeder made elaborate presentation along with the salient features for its utility and the availability of market support system through pulp and plywood industries. Above all, it is presented before the committee that the proposed clone in Melia is amenable for harvest in 24 months for pulp and paper production and 48 to 60 months for plywood production. It has been presented that the variety is highly amenable for incorporation in various Agroforestry system in the state of Tamil Nadu.

Accordingly the SVRC has approved the proposal and released Melia MD 26 as Melia MTP 2. Now the variety is adopted by one of the leading plywood manufactures in the country M/S Century Ply Boards (India) Limited, Chennai and also by the consortium industries like Ambiply panels and doors, Mettupalayam and Ranga ply, Coimbatore through their plywood contract farming system. The industrial linkage ensure quality planting material supply, technological support, assured buyback and facilitation of felling and transport which attracted many farmers towards adoption of this variety in various agroforestry system.

ii. VARIETAL DEVELOPMENTAL PROCESS FOLLOWED FOR KADAM MTP 1

Parentage	:	Selection from Existing population
Breeding method	:	Clonal selection
Breeding Objectives	:	To identify High Yielding and Short Rotation (HYSR) Clone amenable for varied industrial utility

Breeding Approach	:	❖ Preliminary - Species evaluation ❖ Seed source evaluation - 1st generation ❖ Progeny Test – 2nd Generation ❖ Clonal test – 3rd generation ❖ Industrial quality test for wood quality

PROCESS OF GENETIC TESTS AND RECOMMENDATIONS

Preliminary	:	**Species evaluation**
Seed source evaluation	:	1st generation - (Identified elite plus trees and advanced to progeny test based on growth attributes)
Progeny Test	:	2nd Generation - (Identified elite trees, cloned through mini clonal technology and advanced to clonal test)
Clonal test	:	3rd generation - (Final validation has been completed through yield and quality attributes)
Amplified clonal Test	:	The clone is advanced to farmers / Industrial test on a larger scale for validation of yield attributes.

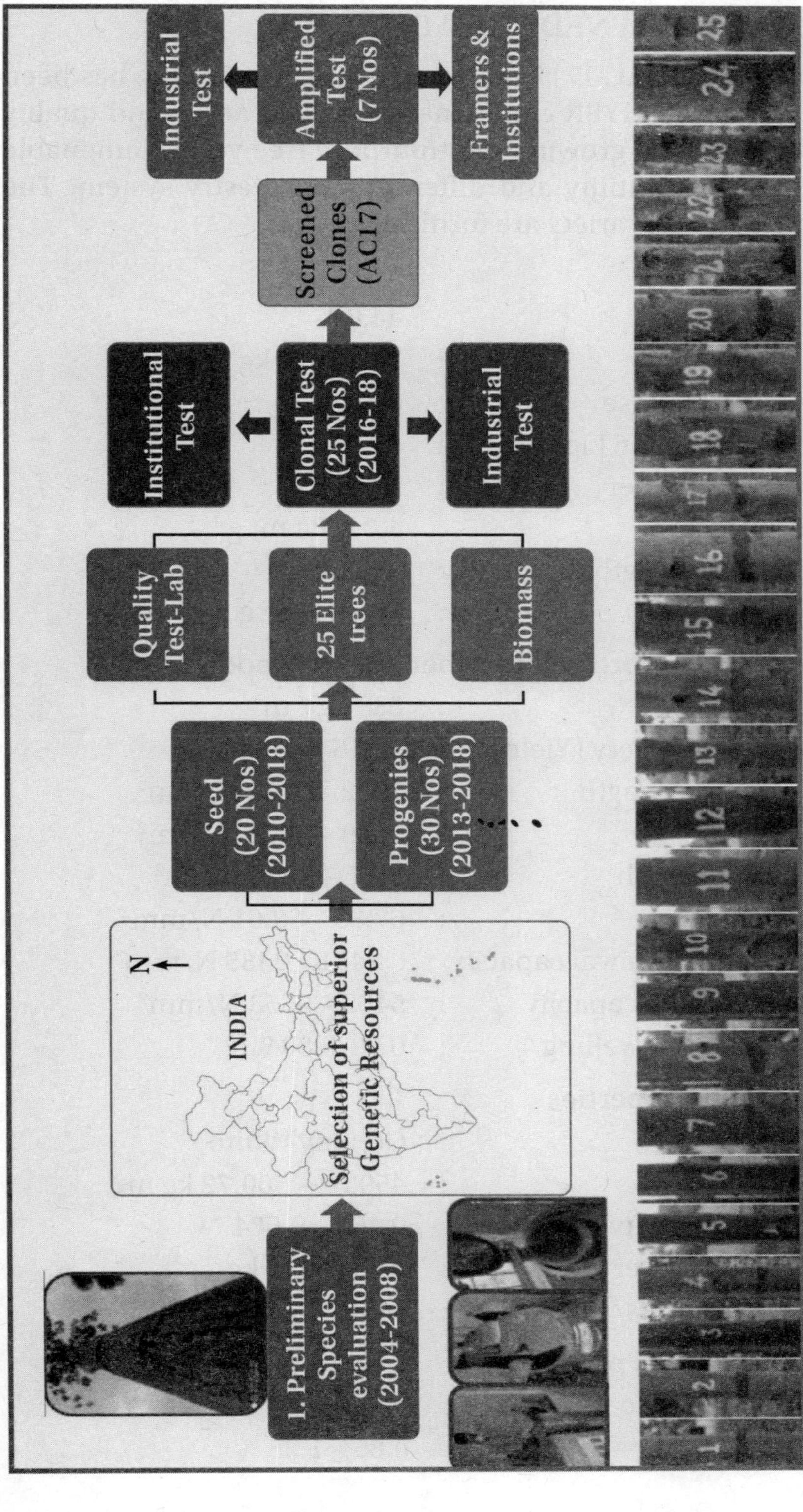

Fig. 16.2. Breeding Plan and Approach

FINAL CLONE SCREENED FOR ADOPTION

A clone (Kadam AC 17) from *Neolamarckia cadamba* has been finally screened as a HYSR clone based on yield and wood quality parameters. It is a fast growing multipurpose tree variety amenable for varied industrial utility and different agroforestry system. The salient features of the variety are furnished below.

a) **Pulp qualities**

- Pulp yield : 44.8 %
- Basic Density : 385 – 415 kg m^{-3}
- Hollo Cellulose : 68.8 %
- Acid Insoluble Lignin : 25.3 %
- Kappa number : 20.4 %
- Burst index : 4.8 K Pa m^2 g^{-1}
- Tensile strength : 74.0 Nm g^{-1}
- Tear index : 6.4 m Nm^2 g^{-1}

b) **Mechanical Properties for timber and Plywood**

- Density : 633 kg / m^3
- Veneer recovery (Yield) : 52.16 %
- Bending strength : 3922 – 4104 N/mm^2
- MOE : 7789 – 7879 N/mm^2
- Shear strength : 2.17 – 2.71 N/mm^2
- MOR : 67.89 – 68.61 N/mm^2
- Screw withdrawal capacity : 2451 – 2485 N/mm^2
- Nail holding capacity : 5462 – 5553 N/mm^2
- Thickeness swelling : 0.79-0.80 %

c) **Match Wood Properties**

- Colour : Creamy White
- Density : 459.62 – 500.72 kg m^{-3}
- Specific gravity : 0.468 – 0.584
- Splints recovery : 32.4 – 39.8 kg
- No. of. Splints/100g : 1450 - 1782

d) **Thermo chemical properties**

- Calorific value : 3850 - 4100 cal/kg
- Ash : 0.85 – 1.25 %

e) **Fodder Quality**

- Moisture Content : 60.70 %
- Dry Matter production : 44.10 %
- Ash Content : 5.91 %
- Crude fat : 1.79 %
- Crude Protein : 16.18 %
- Carbohydrate : 15.60

PROCESS OF VARIETY RELEASE

Based on the various evaluation trails conducted during the last 15 years, a superior clone FCRIAC17 has been placed before the University Varieties and Technologies release screening committee (UVTRSC) for approval on 17.12.2018. The salient features of the variety along with various breeding methods adopted have been presented before the committee and the committee has approved and recommended for submission to state variety release committee.

The variety is then placed before the state variety release committee held on 07.01.2019 chaired by the Agricultural Production Commissioner and Principal Secretary to Government of Tamil Nadu, Agricultural Department in the presence of SVRC Members. The tree breeder made elaborate presentation along with the salient features for its utility and the availability of market support system through pulp and plywood industries. It is presented before the committee that the proposed clone in Kadam is amenable for incorporation in Multifunctional Agroforestry system in the state of Tamil Nadu which will act as a alternate land use system to double the productivity.

Accordingly, the SVRC has approved the proposal and released Kadam FCRIAC 17 as Kadam MTP 1. Now the variety is under mass multiplication and promotional activities are in process. One of the Consortium industry M/S Century ply boards (I) Limited have adopted this variety and incorporated in the contract farming system.

SUMMARY

The growing importance of plantation and agroforestry due to the policy directions and the legal restrictions on felling have complied to develop new varieties in various tree species for incorporation in operational plantation development programme. In India, over 25 different tree species are commercially deployed in forest plantations and agroforestry programme. A few species like poplars, eucalyptus and casuarina wherein significant genetic improvement programme has been carried out which identified high yielding clones and these clones are popularized through agroforestry programme. For other species, which account for almost 50 % of the plantation development activities received very little tree improvement programme and the associated varietal developmental process. Considering this into account Tamil Nadu Agricultural University has prioritized over 20 different tree species and initiated tree breeding and improvement programme through provenances / seed sources / progeny / clone / hybrid development and evaluation programme. Based on these systematic research and developmental process, TNAU has established a standard variety developmental protocol which is unique to tree variety development. Accordingly over 8 different varieties have been developed and released for adoption by the farmers, wood based industries and other stakeholders through State Variety Release Committee. Another 12 varieties representing 8 different species are in various stages of evaluation which may be released in the future years. Unlike agriculture, the varietal developmental process in tree species is long term in nature and requires huge volume of resources in terms of space, time and money. Above all, the breeding programme is highly complex and demand simultaneous evaluation taking into consideration the juvenile superiority of the varieties. Not only the wood and biomass yield is important, but also the wood quality which decides the acceptability of the variety by the consuming industries. Hence Tamil Nadu Agricultural University has conceived and established a strong and systematic tree variety developmental protocol which is furnished in this chapter for two different species. The protocols furnished will act as a great piece of information to the scholars and scientists involved in tree breeding process. The protocol presented can act as a model and the practicing tree breeders can suitably moderate and manipulate the breeding process depending on the species and the breeding objective envisaged.

17

TREE VARIETAL REGISTRATION

INTRODUCTION

The growing importance of Intellectual Property Rights (IPR) in biological system have received greater attention across the world in order to protect the rights of breeders, farmers, researchers and other industries involved in development of plant varieties. Significant developments have been made in agricultural and horticultural sector wherein wide range of crop plants have been protected for their Intellectual Property Rights. However application of Intellectual Property Rights in forestry sector has not received attraction till the recent past. With the growing demand for wood and wood products and the decrease in supply from natural forest have ushered in a total mismatch between demand and supply of wood and wood products. The Government of India has directed all the wood based industries in the country to develop its own raw material resources by establishing suitable linkage with farmers and to promote intensive agroforestry systems. These developments have made several research institutes and wood based industries to develop improved genetic resources for higher yield and short rotation through systematic tree breeding and improvement programme. These approaches have witnessed development of superior genetic stocks/ clones in some of the fast growing tree species like poplars, eucalyptus, casuarina, melia, kadam, sissoo and biofuel crop like Jatropha. The increase intake on plantation forestry and the successful development and release of improved tree varieties have created attraction among the breeders and farmers to protect their varieties for Intellectual Property Rights. To subserve these needs, the Government of India as early in 2001 have enacted Protection of Plant Varieties and

Farmers Right Act (PPVFRA) to and provide legal protection for the newly developed varieties. This besides, the rights of breeders, farmers and others involved in varietal development process are also protected through this act. Hence the current chapter on "Tree Varietal Registration" has been conceived and presented with the Salient features of PPVFRA (2001), DUS testing guidelines, DUS descriptors, DUS testing centres etc., to benefit the graduates for academic purposes; tree breeders to follow the guidelines to protect the varieties and others to know the status of varietal registration and to apply them towards protection of the developed varieties and the proposed varieties.

DEFINITION OF TERMINOLOGIES

Breeder	:	Breeder means a person or group of persons or a farmer or group of farmers or any institution which has bred, evolved or developed any variety;
Essentially Derived Variety	:	Essentially Derived Variety is predominantly derived from such initial variety, or from a variety that itself is predominantly derived from such initial variety, while retaining the expression of the essential characteristics that results from the genotype or combination of genotype of such initial variety.
Extant Variety	:	Extant Variety means a variety available in India which is— • Notified under section 5 of the Seeds Act, 1966 • Farmers' variety; or • A variety about which there is common knowledge; or • Any other variety which is in public domain;
Farmers' Variety	:	Farmers' Variety means a variety which has been traditionally cultivated and evolved by the farmers in their fields; or it is a wild relative or land race of a variety about which the farmers possess the common knowledge.
Propagating Material	:	Propagating Material means any plant or its component or part thereof including an intended seed or seed which is capable of, or suitable for, regeneration into a plant.

Variety : Variety means a plant grouping except micro-organism within a single botanical taxon of the lowest known rank.

- Variety is defined by the expression of the characteristics resulting from a given genotype of that plant grouping;
- Distinguished from any other plant grouping by expression of at least one of the said characteristics; and
- Considered as a unit with regard to its suitability for being propagated, which remains unchanged after such propagation,
- Includes propagating material of such variety, extant variety, transgenic variety, farmers' variety and essentially derived variety.

Sui generis : *Sui generis* means 'the only one of its kind' or 'Self generated'

DUS TESTING – HISTORY, SCOPE AND IMPORTANCE

Plants occurring or developed within a country are invaluable national resources. Farmers contribute in conserving, improving and making available plant genetic resources for developing new plant varieties. Plant breeders play a pioneering role in the development of new varieties. It is important to protect the plant genetic resources and the rights of farmers as well as plant breeders who have contributed for the conservation and development of plant genetic resources.

Originally, India did not have any plant varieties protection law or any system of registration of plant varieties for protection of rights of plant breeders. The only system available was Notification of crop varieties under the Seeds Act, 1966. This was to enable certification of seed quality and assure supply of quality seeds to farmers in right place, right time and right quantity. Indian Patents Act, 1970 excludes agriculture and horticulture methods of production from patentability.

Need for protection of new varieties

- ❖ Incentives to institutions / individuals / organizations engaged in creative work.

- Monetary return on investment in research.
- Plant Breeders' Rights for sustaining plant breeding activity.
- Adequate protection and recognition of farmers as suppliers of germplasm.

Plant Breeder's Rights will help in

- Stimulating research and developing new varieties of plants.
- Facilitating growth of seed industry.
- Ensuring high quality.
- Accelerating agriculture growth.

Considering the importance of protecting the rights of farmers and plant breeders, an inter-governmental organization called **'International Union for Protection of New Varieties of Plants (UPOV)'** was established by the International Convention for Protection of New Varieties of Plants. This convention was adopted in 1961 and came in force in August 1968. UPOV has its headquarters at Geneva.

Under such circumstances, World Trade Organization (WTO) was established in January, 1995 to which India was also a signatory. WTO has at least half a dozen Intergovernmental agreements that directly influence agriculture. Agreement on Agriculture (AoA) is one of the provisions of Trade Related Aspects of Intellectual Property (TRIPs) agreements. Article 27 (3) b of TRIPs agreements envisages that member countries shall protect the new plant varieties through:

- Patent [OR] *Sui generis system* [OR] Combination of both.

India chose to develop its own *Sui generis system* of plant variety protection. *Sui generis* means 'the only one of its kind' or 'Self generated'. The Indian Parliament, in November 2001, passed 'The Protection of Plant Varieties and Farmers' Rights Act (PPV FRA)'. It is a *Sui generis* law for protection of plant varieties, developed by integrating the rights of breeders, farmers and village communities, also taking care of the concerns of equitable sharing of benefits. It is in conformity with UPOV Convention (1978) Act.

Intellectual Property Rights (IPR) means a right which is derived from itself, or which is not a part of the patent system. The

Act introduces important approaches to farmers' rights and has **adapted some provisions of UPOV** to the Indian context. PPV FRA of India is the only Legislation for PVP in the whole world, which incorporates the Farmers' Rights as an important element.

SALIENT FEATURES OF PPVFR ACT, 2001

Registrable plant varieties

As per the provisions of PPVFR Act, the following four varieties are amenable for registration

i. **New Varieties**
   - It includes a variety which is not in public domain in India earlier than one year before the date of filling.
   - For outside India, It is earlier than four years and earlier than six years in case of trees or Vines.

ii. **Extant Variety (EV)**
   - It includes a variety that is notified under the provisions of seed act 1996
   - It also includes a variety about which a common knowledge or a farmers' variety or any other variety which is in public domain is considered as an extant variety

iii. **Farmers Variety (FV)**
   - It includes the variety which has been traditionally cultivated and evolved by the farmers in their fields.
   - It also includes a variety which is a wild relative or land race of a variety about which farmers have common knowledge.

iv. **Essentially Derived Variety (EDV)**
   - It is a essentially deriver variety from such initial variety when it is predominantly derived form such initial variety or from a variety that itself is predominantly derived from such initial variety, while retaining the expression of the essential characteristics that result from the genotype or combination of genotype of such initial variety and it is clearly distinguishable from such initial variety.

- This EDV conforms to such initial variety from the genotype or combination of genotype of such initial variety.

Criteria for registration of tree varieties

The tree varieties intended for registration leading to protection should have the following criteria which form eligibility criteria for tree variety registration.

- Novelty
- Distinctness
- Uniformity
- Stability

Distinctness

Must be clearly distinguishable for at least one essential character from the varieties of common knowledge in India and outside. Essential Character is a heritable trait which is determined by one or more genes or other heritable determinants that contribute to the principal features, performance or value of the plant variety.

Novelty

Variety was not placed for scale for a period exceeding 12 months on the date of application. A period of four years in the case of annual crop varieties and six years for trees and vines since first registration outside India on the date of registration in India.

Uniformity

Uniform, subjected to the variation that may be expected and allowed due to the specific nature of the reproduction of the crop such as vegetative, self and cross pollinated.

Stability

Stable when all essential characters remain unchanged after repeated propagation or a specified cycle of propagation.

Persons eligible for varietal registration

The following persons are eligible to register the plant variety

- Any breeder of the variety

- Any successor of the breeder
- Any person being the assignee of the breeder
- Any farmer who evolved the variety
- Any group of farmers or community of farmers claiming to be the breeder of the variety
- Any person authorized on behalf of farmers
- University evolved the variety
- Public funded Agricultural Institutions claiming to be breeder of the variety

DUS TESTING PROCEDURE

i. Application (Technical Questionnaire)

The breeder seeking protection for his new variety should submit an Application Form and detailed technical questionnaire. The breeder should describe the characteristics of his new variety to claim registration.

ii. Seed of a candidate variety

The breeder is required to supply prescribed quantities of seed or other reproductive material when required for sowing of planting. The growing tests are normally conducted for a minimum of two successive growing seasons under the control of authority.

The layout of the tests, their location and duration, the numbers of centers and nature of the observations to be made on the plants are all prescribed in regulations and working rules laid down by the authority.

iii. Reference collection

In order to establish distinctness, it is usually necessary to compare the candidate variety with existing varieties. For this purpose, the test centers maintain reference collections of authentic samples of seed of existing varieties. The candidate varieties and appropriate varieties from the reference collection are compared in the field and laboratory tests. Normally comparisons are limited to groups or varieties, which are similar to the candidate variety.

iv. Number of years of testing

Tests of DUS normally take two years for self-pollinated crops but cross pollinated crops may require three years.

v. Characters used for DUS

The distinctness of self-pollinated crops can be established using characters which can be assessed by visual examination and whose expression falls into clearly defined discrete states. Because of the genetic diversity of cross pollinated crops, many of the varietal characteristics are on a continuous scale of expression and require measurement. Usually, distinctness can be determined only on the basis of statistical analysis in cross pollinated crops.

vi. Number of characters

The number of characters to be evaluated depends on (i) the number or readily identifiable characters associated with the crop (ii) the amount of effort required recording those characters and (iii) the value of those characters in distinguishing varieties. It is suggested that the number of characters should be kept as low as possible.

To register a variety under PVP, it should pass DUS test for implementation of breeder and farmers rights. Based on DUS test, a varietal description is produced, which is often used for field inspection to have certification / self certification.

TREE DUS TESTING CENTRES

S.No	Institute / Organisation	Prioritized species
1.	Institute of Forest Genetics and Tree Breeding, Coimbatore, Tamil Nadu.	i. Casuarina (Casuarina species) ii. Eucalyptus (Eucalyptus species) iii. Melia (*Melia dubia*)
2.	Forest College and Research Institute, Tamil Nadu Agricultural University, Mettupalayam, Tamil Nadu.	i. Jatropha (*Jatropha curcas*) ii. Neem (*Azadirachta indica*) iii. Karanj (*Pongamia pinnata*)

S.No	Institute / Organisation	Prioritized species
3.	Department of Tree Improvement & Genetic Resources Dr. Y.S. Parmar University of Horticulture & Forestry Nauni-173230, Solan, Himachal Pradesh	i. Poplar (*Populus deltoides*) ii. Willow (Salix species)
4.	Himalayan Forest Research Institute ICFRE, Conifer Campus Shimla-171013, Himachal Pradesh	i. Deodar (*Cedrus deodara*) ii. Chir Pine (*Pinus roxburghii*)

TREE DUS DESCRIPTORS

i. Approved DUS Descriptors for various species (Adopted from www. plantauthority.gov.in)

1. CASUARINA (Casuarina species)

S.No	Characteristics	Descriptors
1.	Branchlet: Colour	Light green
		Dark green
		Bluish green
		Yellowish green
2.	Branchlet: Length	Short (<15 cm)
		Intermediate (15-25 cm)
		Long (>25 cm)
3.	Branchlet: Thickness	Thin (<0.3 mm)
		Intermediate (0.3- 0.6 mm)
		Thick (>0.6 mm)
4.	Branchlet: Leaves per node	Less than 6 leaves
		6 leaves
		7 leaves
		8 leaves
		9 leaves
		10 leaves
		More than 10 leaves

S.No	Characteristics	Descriptors
5.	Branchlet: Internode length	Short (< 5 mm)
		Medium (5- 8mm)
		Long (>8mm)
6.	Anthocyanin in the leaf tip	Absent
		Weak
		Medium
		Strong
7.	Pubescence in leaf tips	Absent
		Present
8.	Leaf tip marks	Absent
		Present
9.	Pruning Scars	Isosceles triangular
		Equilateral triangular
		Eye-shaped
10.	Lenticels: Grouping	Individual
		Vertical lines
		Slanting / Curved lines
		Cluster
		Mixed
11.	Lenticels: Density	Very Low (<20 per sq. cm)
		Low (21-30per sq. cm)
		Intermediate (31-40 per sq. cm)
		High (41-50per sq. cm)
		Very High (>50 per sq. cm)
12.	Lenticel: Shape	Round
		Oval
		Eye shape
13.	Lenticel: Size	Small (<0.5mm)
		Medium (0.5-1.5 mm)
		Large (>1.5mm)
14.	Crown: Shape	Lanceolate
		Conical
		Columnar

S.No	Characteristics	Descriptors
15.	Bark: Texture	Smooth
		Rough
		Very rough
16.	Bark:Colour	Grey
		Purple
		Pinkish purple
		Brown
17.	Protrusion on main stem	Absent
		Present
18.	Branching pattern	Single
		Paired
		Cluster
19.	Branch: Angle	Upward (<60°)
		Horizontal (60to 90°)
20.	Branch: Thickness	Thin (<1.5 cm)
		Medium (1.5- 2.5 cm)
		Thick (>2.5 cm)
21.	Protrusions on primary branches	Absent
		Present
22.	Bearing of deciduous branchlets	Straight
		Drooping
23.	Sexual system	Not flowering
		Dioecious Male
		DioeciousFemale
		Monoecious (Male + Female)
24.	Male Inflorescence: Position	Terminal
		Axillary
		Mixed
25.	Male inflorescence: Length	Short (<1.5 cm)
		Medium(1.6 to 3.5 cm)
		Long (3.6 to 5.5cm)
		Very long (>5.5cm)

S.No	Characteristics	Descriptors
26.	Male inflorescence: flowers per whorl	Less than 6 flowers
		6 flowers
		7 flowers
		8 flowers
		9 flowers
		10 flowers
		More than 10
27.	Male flower: Anther colour	Cream
		Yellow
		Pink
		Red
28.	Cone: Length	Short (<1 cm)
		Intermediate (1.1 to 2.5 cm)
		Long (>2.5 cm)
29.	Cone: Shape	Round
		Cylindrical
		Ovate
30.	Cone:Breadth	Narrow (<1 cm)
		Intermediate (1-2cm)
		Broad (>2cm)
31.	Cone: Valves per whorl	Up to 6 valves
		7 valves
		8 valves
		9 valves
		10 valves
		More than 10
32.	Cone: Valve texture	Spiny
		Blunt
33.	Cone: Pedicel length	Short (<1 cm)
		Intermediate (1- 2.5 cm)
		Long (>2.5 cm)
34.	Cone: Pedicel shape	Straight
		Curved

2. EUCALYPTUS (Eucalyptus species)

S.No	Characteristics	Descriptors
1.	Juvenile leaf: Shape	Lanceolate
		Ovate
2.	Juvenile leaf: Anthocyanin colouration	Absent or weak
		Medium
		Strong
3.	Mature leaf: Shape	Narrow Lanceolate
		Lanceolate
		Ovate
4.	Mature leaf: Base Symmetry	Symmetric
		Asymmetric
5.	Mature leaf: Base Shape	Obtuse
		Cuneate
		Attenuate
6.	Mature leaf: Apex Shape	Acute
		Obtuse
		Subulate
7.	Mature leaf: Blade area	Small (<20.0 cm^2)
		Medium(20.0 to 46.0 cm^2)
		Large (>46.0 cm^2)
8.	Mature leaf: Length	Short (<12.0 cm)
		Medium (12.0 to 21.0 cm)
		Long (>21.0 cm)
9.	Mature leaf: Breadth	Narrow (<2.5 cm)
		Medium (2.5 to 4.0 cm)
		Wide (>4.0 cm)
10.	Mature leaf: Petiole length	Short (<2.0 cm)
		Intermediate (2.0-3.0 cm)
		Long (>3.0 cm)
11.	Mature leaf: Blade ratio	Slightly elongated (<4.0 times long leaf)
		Moderately elongated (4-6 times long leaf)
		Very elongated (>6 times long leaf)

S.No	Characteristics	Descriptors
12.	Mature leaf: Waxiness of upper side	Absent or weak
		Medium
		Strong
13.	Trunk: Clear bole Height	Below lower 1/3rd height
		With in middle 1/3rd height
		Above top 1/3 rd height
14.	Crown: Shape	Lanceolate
		Conical
		Columnar
15.	Primary branch: Scar type	Open
		Close
16.	Primary branch: Scar shape	Inverted "V"
		Spherical
17.	Primary branch: Scar periphery projection	Downward
		Horizontal
		Flat
		Depressed
18.	Primary branch: Self pruning	Present
		Absent
19.	Primary branch: Thickness	Small (<1/8 th of main stem)
		Medium (1/8th -1/4th of main stem)
		Thick (>1/4 th of main stem)
20.	Primary branch: Attitude	Upward
		Horizontal
		Drooping
21.	Bark: Texture	Rough
		Smooth
22.	Bark: Annual Peeling	Absent
		Present
23.	Bark: Peeling type	Strip
		Flake
		Mixed

S.No	Characteristics	Descriptors
24.	Bark: Colour of fresh bark	Light brown
		Light green
		Light grey
25.	Bark: Colour of dried bark	Light brown
		Brown
		Grey
26.	Bark: Colour of rhytidome bark	Light brown
		Dark brown
		Grey
27.	Trunk: Waxiness (excluding rhytidome)	Absent or weak
		Medium
		Strong
28.	Flower: Operculum shape	Hemispherical apiculate
		Elongated
		Conical
29.	Fruit: Base shape	Conical
		Spherical
		Hemispherical
30.	Fruit: Operculum scar	Absent
		Present
31.	Fruit: Pedicel length	Short (<4.5 mm)
		Medium (4.5 to 6.5 mm)
		Long (>6.5 mm)
32.	Fruit: Peduncle length	Short (<1.0 cm)
		Medium (1.0 to 1.5 cm)
		Long (>1.5 cm)
33.	Fruit: Width	Narrow (<5 mm)
		Medium (5-6 mm)
		Broad (>6 mm)

3. NEEM (Azadirachta indica)

The DUS characteristics of Neem are provided below

S.No	Characteristics	Descriptors
1.	Tree habit	Spreading
		Semi-upright
		Upright
2.	Leaflet length	Short (<5.0 cm)
		Medium (5.0 -10 cm)
		Long (>10.0 cm)
3.	Leaflet width	Narrow (<1.5 cm)
		Medium (1.5-3.0 cm)
		Broad (>3.0 cm)
4.	Petiole: Length	Very short (<4.0 cm)
		Short (4.1 - 8.0 cm)
		Medium (8.1-12.0 cm)
		Long (>12.0 cm)
5.	Inter Leaflet: Rachis Length	Very short (<2.0 cm)
		Short (2.0 - 3.0 cm)
		Medium (3.1- 4.0 cm)
		Long (4.1 - 5.0 cm)
		Very long (>5.0 cm)
6.	Leaflet shape	Falcate
		Lanceolate
7.	Leaf margin: Serration	Serrate
		Dentate
8.	Petals shape	Spatulate
		Obovate
9.	Fruit length	Short (<1.5 cm)
		Medium (1.5-2.0 cm)
		Long (>2.0 cm)
10.	Fruit width	Narrow (<1.0 cm)
		Medium (1.0-1.5 cm)
		Broad (>1.5 cm)

S.No	Characteristics	Descriptors
11.	Fruit shape	Oblong
		Oval
		Globular
12.	Fruit: Apex	Round
		Acute
13.	Fruit: Stalk cavity	Absent
		Present
14.	Presence of neck	Absent
		Present
15.	Presence of fruit shoulder	Absent
		Present
16.	Seed length	Short (<1.0 cm)
		Medium (1.0-1.5 cm)
		Long (>1.5 cm)
17.	Seed width	Narrow (<0.5 cm)
		Medium (0.5-0.8 cm)
		Broad (>0.8 cm)
18.	Seed shape	Oblong
		Ovate
19.	Seed tip shape	Acute
		Obtuse
20.	Oil content	Low (<36%)
		Medium (36% - 41%)
		High (41% - 45%)
		Very High (> 45%)
21.	Azadirachtin content	Low (<0.30%)
		Medium (0.30% -0.60%)
		High (>0.60%)

4. JATROPHA (*Jatropha curcas*)

The DUS trait of Jatropha has been provided in the following table.

S.No	Characteristics	Descriptors
1. Plant character		
1a.	Plant height (cm)	Short (<130.0)
		Medium (130.0 - 200.0)
		Tall (>200.0)
1b.	Growth habit	Erect
		Bushy
		Spreading
2. Stem character		
2a.	Stem bulginess	Absent
		Present
2b.	Young shoot colour	Green
		Pinkish red
3. Leaf character		
3a.	Leaf length (cm)	Short (<8.0)
		Medium (8.0 - 15.0)
		Long (>15.0)
3b.	Leaf width (cm)	Narrow (<7.0)
		Medium (7.0 - 14.0)
		Broad (>14.0)
3c.	Petiole length (cm)	Short (<10.0)
		Medium (10.0 - 16.0)
		Long (>16.0)
3d.	Leaf shape	Cordate
		Palmate
3e.	Leaf texture	Smooth
		Coarse
		Rough
		Leathery
3f.	Petiole colour	Greenish pink
		Green

S.No	Characteristics	Descriptors
3g.	Juvenile leaf colour	Dark pink
		Greenish pink
4. Fruit character		
4a.	Fruit length (cm)	Short (<2.5)
		Medium (2.5 - 3.5)
		Long (>3.5 cm)
4b.	Fruit width (cm)	Narrow (<2.1)
		Medium (2.1 - 2.9)
		Broad (>2.9)
4c.	Fruit shape	Oval
		Spherical
		Oblong
4d.	Fruit: Prominence of ridges	Weak
		Medium
		Strong
4e.	Fruit: Presence of shoulder	Absent
		Present
4f.	Fruit: Presence of segment	Absent
		Present
4g.	Fruit colour	Dark yellow
		Yellowish orange
		Reddish orange
5. Seed character		
5a.	Seed length (cm)	Short (<1.5)
		Medium (1.5 - 2.0)
		Long (>2.0)
5b.	Seed width (cm)	Narrow (<0.60)
		Medium (0.60 - 1.10)
		Broad (>1.10)
5c.	Seed shape	Ellipsoid
		Ovate

S.No	Characteristics	Descriptors
5d.	Seed carnicle	Big
		Medium
		Small
5e.	Seed coat : Kernel ratio	High (> 65 %)

5. KARANJ (*Pongamia pinnata*)

S.No	Characteristics	Descriptors
1.	Tree habit	Semi-upright
		Upright
		Drooping
2.	Stem type	Single stem
		Multi stem
3.	Leaflet: Length	Short (< 6 cm)
		Medium (6-12 cm)
		Long (> 12 cm)
4.	Leaflet: Width	Narrow (< 3 cm)
		Medium (3-6 cm)
		Broad (> 6 cm)
5.	Petiole length	Short (<3.0 cm)
		Medium (3.0-6.0 cm)
6.	Inter leaflet: Rachis length	Long (>6.0 cm)
		Short (<3.0 cm)
		Medium (3.0- 5.0 cm)
		Long (>5.0 cm)
7.	Leaflet shape	Ovate
		Elliptical
8.	Terminal leaflet: Shape	Deltoid
		Orbiculate
		Lanceolate
		Obovate
		Elliptic
		Ovate

S.No	Characteristics	Descriptors
9.	Terminal leaflet: Apex	Acute
		Acuminate
		Cuspidate
		Mucronate
10.	Terminal leaflet: Base	Cuneate
		Oblique
		Rounded
		Truncate
11.	Leaflet: No. of primary veins	Sparse (>5)
		Medium Dense (5-7)
		Dense (>7)
12.	Flower colour	Pinkish white
		Whitish Yellow
13.	Pod length	Short (< 3 cm)
		Medium (3-6 cm)
		Long (> 6 cm)
14.	Pod width	Narrow (< 1.8 cm)
		Medium (1.8-2.5 cm)
		Broad (> 2.5 cm)
15.	Pod colour	Brown
		Yellowish grey
16.	Pod flatness	Flat
		Slightly swollen
		Swollen
17.	Pod shape	Elliptic
		Oblong
18.	Pod tip: Curvature of beak	Curved
		Slightly curved
		Straight
19.	Pod margin	Convex
		Concave

S.No	Characteristics	Descriptors
20.	Seed length	Short (< 1.8 cm)
		Medium (1.8-2.5 cm)
		Long (> 2.5 cm)
21.	Seed width	Narrow (< 1.0 cm)
		Medium (1.0-1.5 cm)
		Broad (> 1.5 cm)
22.	Seed colour	Reddish brown
		Light Brown
		Brown
23.	Seed shape	Ovate
		Oblong
		Reniform

6. MELIA (*Melia dubia*)

S.No	Characteristic	Descriptors
1.	Stem Scar: Shape	Heart shaped
		Elliptic
2.	Stem scar: Length (cm)	Short (<1.0)
		Medium (1.0 to 3.0)
		Long (> 3.0)
3.	Stem Scar: Width(cm)	Narrow (<0.5)
		Medium (0.5 - 1.25)
		Broad (> 1.25)
4.	Bark : Colour	Light green
		Dark green
		Light Brown
		Dark brown
		Purple
5.	Leaf: Waxiness	Absent
		Present
6.	Stem lenticels: Appearance	Flat
		Partially embossed 2
		Embossed

S.No	Characteristic	Descriptors
7.	Stem Lenticels: Pattern	Scattered
		Linear
8.	Stem Lenticels: Density (cells/sq cm)	Low (<5)
		Medium (5-10)
		High (> 10)
9.	Stem Lenticels: Shape	Linear
		Star
		Ellipsoidal
		Round
		Mixed
10.	Stem Lenticels: Size (mm)	Small (<2)
		Medium (2-7)
		Big (>7)
11.	Bark: Peeling	Absent
		Present
12.	Bark: Peeled Surface	Smooth
		Rough
13.	Rachis: Attitude	Upright
		Horizontal
		Drooping
14.	Rachis: Length(cm)	Short (< 15)
		Medium (15-30)
		Long (>30)
15.	Rachis: Number of leaflets	<8 pairs
		8-12 pairs
		> 12 pairs
16.	Leaflet: Shape	Lanceolate
		Elliptic
		Ovate
		Cordate

S.No	Characteristic	Descriptors
17.	Leaflet: Margin	Entire
		Crenate
		Serrate
		Dentate
		Mixed
18.	Leaflet: Base	Oblique
		Partially Oblique
		Symmetric
19.	Leaflet Apex: Shape	Acute
		Acuminate
		Caudate
20.	Leaflet Lamina: Length (cm)	Short (<5)
		Medium (5- 8)
		Long (>8)
21.	Leaflet: Breadth(cm)	Narrow (<2)
		Medium (2-3.5)
		Broad (>3.5)
22.	Rachis: Anthocyanin Colouration	Absent
		Present
23.	Petiolule: Length(cm)	Short (<0.5)
		Medium (0.5 to 1)
		Long (>1)
24.	Branch: Attitude	Upright
		Horizontal
		Drooping
25.	Stem: Straightness	Not Straight
		Not Straight
26.	Stem: Clear Bole Height (Without treatment)	<50 % of tree height
		50 to 70 % of tree height
		>70% of tree height

S.No	Characteristic	Descriptors
27.	Branch: Thickness	Small: <1/8 of Stem Diameter
		Medium: 1/8- ¼ of Stem Diameter
		Thick: >1/4 of Stem Diameter
28.	Crown: Shape	Spherical
		Hemispherical
29.	Inflorescence: Compactness	Sparse
		Dense
30.	Inflorescence: Length (cm)	Short (< 10)
		Medium (10-20)
		Long >20)
31.	Inflorescence: Peduncle Length(cm)	Short (< 3)
		Long (Above 3)
32.	Flowers: Merosity	Tetramerous
		Pentamerous
		Hexamerous
		Septamerous
33.	Staminal Tube:Shape	Conical
		Cylindrical
34.	Pistil: Length (mm)	Short (≤8)
		Long (Above 8)
35.	Fruit: Length (cm)	Short (< 3)
		Medium (3-5)
		Long (>5)
36.	Fruit: Width (cm)	Narrow (<2)
		Medium (2-3.5)
		Broad (>3.5)
37.	Fruit: Shape	Globular
		Ovoid
		Ellipsoid

S.No	Characteristic	Descriptors
38.	Endocarp: Length (cm)	Short (< 2)
		Medium (2-3.5)
39.	Endocarp: Width (cm)	Narrow (<1.5)
		Medium (1.5-2)
		Broad (>2)
40.	Endocarp: Shape	Globular
		Ovoid
		Ellipsoid

7. POPLAR (*Populus deltoides*)

The characteristics described in the following table are used for testing of varieties for their DUS in poplar.

S.No	Characteristics	Descriptors
1.	Leaf: Shape	Cordate
		Deltoid
2.	Leaf: Waviness	Undulation
		Flat Surface
3.	Leaf: Anthocyanin colour on mid rib	Absent/Weak
		Present
4	Leaf: Distribution of Anthocyanin colour on mid rib	Base
		Base-Middle
		Whole Midrib
5	Leaf: Shape of Tip	Narrow Acute
		Narrow Long Acuminate
		Broad Long Acuminate
		Narrow Short Acuminate
6	Leaf: Shape of Base	Weakly Cordate
		Medium Cordate
		Strongly Cordate
7	Leaf: lobes	Absent
		Present

S.No	Characteristics	Descriptors
8.	Leaf: Leaf Blade Maximum width (cm)	Small (<20)
		Large (>20)
9.	Leaf: Mid rib length (cm)	Short (<20)
		Long (>20)
10.	Leaf: Petiole length (cm)	Short (<12)
		Long (>12)
11.	Leaf: Total length (cm)	Short (<30)
		Medium (30-32)
		Long (>32)
12.	Leaf: Ratio of length of petiole to length of mid rib	Small (<0.60)
		Medium (0.60-0.70)
		Large (>0.70)
13.	Leaf: Ratio of length of mid rib to maximum width of lamina	Small (<1)
		Large (>1)
14.	Leaf: Angle between midrib and second lower lateral vein(0)	Small (<70)
		Large (>70)
15.	Leaf: Apex Angle(0)	Small (<70)
		Large (>70)
16.	Leaf : Leaf Area (cm)2	Small (upto 270)
		Medium(270-350)
		Large(>350)
17.	Petiole: Colour	Absent
		Present
18.	Branch : Ribs on Branch	Smooth
		Weakly Ribbed
		Ribbed
19.	Branch: Attitude	Curved Up
		Horizontal
		Curved Down

S.No	Characteristics	Descriptors
20.	Stem : Length of Bud (cm)	Small(<0.50)
		Medium (0.50-1)
		Large (>1)
21.	Stem: Presence of Ribs below leaf scar	Round
		Slightly angular
		Angular
		Winged
22.	Stem: Basal Stem Diameter (mm)	Small (<25)
		Large (>25)

8. WILLOW (Salix species)

S.No	Characteristic	Descriptors
1.	Main stem: attitude	Slightly curved
		Strongly curved
2.	Main stem: hairiness	Weak
		Strong
3.	Main stem: number of lenticels/kcm on all sides of stem	Few (<3)
		Medium (3-5)
		More (>5)
4.	Main stem: hairiness of leaf bud	Weak
		Strong
5.	Main stem: number of branches longer than 5 cm	Few (<20)
		Many (>20)
6.	Branch: angle between first 5 cm of branch and main stem	Small (<50)
		Large (>50)
7.	Branch: attitude	Curved upwards
8.	Leaf: Total length (cm)	Short (<12)
		Medium (12-18)
		Long (>18)
9.	Leaf : length of Midrib (cm)	Short (<10)
		Medium (10-15)
		Long (>15)

S.No	Characteristic	Descriptors
10.	Leaf : maximum width (cm)	Narrow (<2)
		Medium (2 - 4)
		Broad (>4)
11.	Leaf : position of maximum width	Below middle
		Approximately at middle
		Above middle
12.	Leaf : shape of base	Acute
		Rounded
13.	Leaf : hairiness of upper side	Weak
		Strong
14.	Leaf : hairiness of lower side	Weak
		Medium
		Strong
15.	Leaf: Number of teeth per cm on the leaf margins at middle of leaf blade	Few (<4)
		Many (>4)
16.	Leaf: Ratio of length of petiole to length of mid rib	Small (<0.07)
		Medium (0.07-0.10)
		Large (>0.10)
17.	Leaf: Ratio of mid rib length to maximum width of lamina length	Small (<8)
		Large (>8)
18.	Leaf:Petiole length (cm)	Short (<1)
		Medium (1-2)
		Long (>2)
19.	Stipule type: shape and orientation	Vertically lanceolate diverging
		Vertically lanceolate converging
		Horizontally Semi-circular spreading
20.	Main stem: Plant height (cm)	Short (<125)
		Medium (125 -250)
		Tall (>250)

S.No	Characteristic	Descriptors
21.	Main Stem: Basal diameter (mm)	Small (<15)
		Large (>15)
22.	Main stem: Inter nodal Length (cm)	Short (<4)
		Medium (4-6)
		Long (>6)
23.	Main stem: Leaf fall (50%)	Early
		Late
24.	Main stem: Leaf emergence	Early
		Late

9. DEODAR (*Cedrus deodara*)

S.No	Characteristics	Descriptors
1.	Tree height (m)	Small (< 20)
		Medium (20-30)
		Tall (>30)
2.	Diameter at breast height (cm)	Small (< 40)
		Medium (40-60)
		Large (>60)
3.	Clear bole length (m)	Short (< 10)
		Medium (10-15)
		Tall (15-20)
		Very Tall (> 20)
4.	Self Pruning ability(m)	Low (up to10)
		Medium (10-20)
		High (>20)
5.	Bark colour	Grey
		Dark Grey
		Greyish Brown
		Reddish Brown
		Brown

6.	Bark pattern	Longitudinal shallow furrows (Moist zone)
		Longitudinal deep furrows (Drier zone)
		Longitudinal smooth cracks
		Smooth shallow cracks
		Smooth
7.	Branch : attitude	Horizontal
		Angular
		Drooping
8.	Crown shape	Conical
		Flattened
9.	Needle: attitude	Erect
		Drooping
10.	Needle colour	Light Green
		Green
		Dark Green
		Silvery Green
		Bluish Green
11.	Chlorophyll content (mg/g)	Low (<1)
		Medium (1-1.5)
		High (>1.5)
12.	Cone length (cm)	Small (<7)
		Medium (7-9)
		Large (>9)
13.	Cone width (cm)	Small (<6)
		Medium (6-7)
		Large (> 7)
14.	Cone length: breadth ratio	Small (<1.25)
		Medium (1.25-1.50)
		Large(>1.50)

15.	No. of seeds per cone	<140
		140-150
		>150
16.	Bending: Maximum load	Low (<100 kgf)
		Medium (100-200 kgf)
		High (>200 Kgf)
17.	Strength	Low (<200 kgf)
		Medium (200-400 Kgf)
		High (>400 Kgf)

10. CHIR PINE (*Pinus roxburghii*)

S.No	Characteristics	Descriptors
1.	Tree height (m)	Small (up to 20)
		Medium (21-30)
		Tall (>30)
2.	Diameter at breast height (cm)	Small (up to 40)
		Medium (40-60)
		Large (>60)
3.	Clear bole length (m)	Short (< 10)
		Medium (10-15)
		Tall (16-20)
		Very Tall (>20)
4.	Self Pruning ability(m)	Low (up to 10)
		Medium (10-20)
		High (>20)
5.	Bark colour	Grey
		Pinkish Brown
		Orange Brown
		Dark Brown
6.	Bark pattern	Smooth, shallow fissures
		Small, deep fissures
		Longitudinal shallow fissures
		Longitudinal deep fissure

7.	Bark thickness (cm)	Thin (<0.5)
		Moderate (0.5 - 1.5)
		Thick (>1.5)
8.	Branch : attitude	Horizontal
		Angular
9.	Crown shape	Conical
		Round
		Umbrella
10.	Needle length (cm)	Small (up to 20)
		Medium (20-35)
		Large (>30)
11.	Needle colour	Light Green
		Green
		Yellowish green
		Dark Green
12.	Chlorophyll content (mg/g)	Low (< 1)
		Medium (1-1.5)
		High (> 1.5)
13.	Cone length (cm)	Small (<10)
		Medium (10-15)
		Large (>15)
14.	Cone width (cm)	Small (<5)
		Medium (6-8)
		Large (>8)
15.	Cone length: breadth ratio	Small
		Medium

ii. DUS traits for other species

1. Kadam (*Neolamarckia cadamba*)

S.No	Characteristics	Descriptors
1.	Bark colour	Brown
		Light Brown
		Black
2.	Bark Smoothness	Smooth
		Moderate
		Rough
3.	Leaf Shape	Elliptic
		Ovate
		Broadly ovate
		Nearly round
4.	Leaf Base Shape	Cordate
		Obtuse
		Acute
5.	Leaf Apex Shape	Acute
		Acuminate
		Apiculate
		Cuspidate
		Obtuse
6.	Leaf Length	Short (< 12 cm)
		Medium (12-21 cm)
		Long (> 21 cm)
7.	Leaf Petiole Length	Short (< 2 cm)
		Intermediate (2-3 cm)
		Wide (> 3 cm)
8.	Leaf Waxiness	Weak (or) Absent
		Medium
		Strong

DURATION OF THE PROTECTION OF TREE VARIETIES

As per the provisions of Section 24 (6) of PPV FR Act, 2001, the duration of protection of tree varieties is furnished below.

- Trees and Vines : 18 Years (Initial grant for 9 years)
- Other than Trees and Vines : 15 Years (Initial grant for 6 years)

SUMMARY

Protection of developed or bred tree varieties is gaining significant importance in the era of Intellectual Property Rights. Unlike agriculture, the varietal registration in tree crop is of recent origin but received attraction among the tree breeders and wood based industries due to the increasing attention on promotion of industrial based forestry and agroforestry plantations. Several research institutes of ICFRE, ICAR and State Agricultural Universities coupled with wide range of wood based industries like timber, pulp and paper, plywood, match splints, agricultural implements, dendropower industries and biofuel sector are involved in varietal developmental process. These developments in tree breeding and improvement have attracted all these agencies towards protection of tree varieties. The PPVFR Act (2001) of Government of India has made a provision for the legal protection of new varieties developed besides protecting the rights of breeders and farmers. Based on this act, several institutes were identified to develop DUS traits for various species. This besides, several other institutes are also involved in development of DUS traits for wide range of species amenable for incorporation in Industrial agroforestry. All these developments and provisions were elaborated in this chapter which will act as a ready recokner for the tree breeders and farmers for protection of varieties and graduating scholars towards meeting the academic and competitive needs.

18

TREE BREEDING INSTITUTIONS AND ORGANIZATIONS

INTRODUCTION

The forest research in general and tree improvement research in particular is aimed to contribute to the nations well being in terms of wood and wood products supply besides catering to the needs of environmental and ecosystem stability. This ability of forest and forest product research needs research organizations and institutions with well organized structures, well defined objective, effective management and help to achieve high standards of performance. The institutions and organizations should establish focused research leading to development of varieties and technologies. This may lead to increased potential for commercializing the technologies to benefit the country in general and the society in particular. Among various sectors of forest and forest product research, tree breeding and improvement forms a core area of research which may lead to production of high yielding varieties, short rotation varieties and tree varieties amenable for varied domestic and industrial utility coupled with resistance towards biotic and abiotic factors. Wide range of institutions, organizations and wood based industries are involved in tree breeding and improvement programme which contributed significantly towards productivity and wood quality improvement. Several research institutes and paper industries in Brazil have made tremendous improvement in eucalyptus through systematic and organized tree breeding and improvement strategies. Similarly organized tree breeding by various national and international institutions and organizations resulted in productivity improvement in poplars, pines, eucalyptus, casuarina, salix, bamboos, paulownia, acacia hybrid etc., Hence the current chapter has been designed to collect and furnish the institutions and organizations involved in tree breeding and improvement programme both nationally and globally.

INSTITUTIONS INVOLED IN TREE BREEDING AND IMPROVEMENT

1. INTERNATIONAL INSTITUTES

i. Kenya Forest Research Institute, Kenya

Kenya Forest Research Institute is a research institute and incorporated as a state corporation and established in the year 1986 towards carrying out research in forestry and allied natural resources. The major area of tree breeding research include - breeding programme for Pinus and Cupressus, the development of propagation technology and establishment of seed orchards.

ii. Forestry Research Institute of Ghana

This institute was established as a research unit in the year 1962 and then the research institute is pronounced as Forest Product Research Institute under the Ghana Academy of Sciences in the year 1964. The institute was placed under the Council for Scientific and Industrial Research in the year 1968 and mandated to establish centre of excellence in forestry research to cater to the needs of humid tropical species. The major tree breeding objectives of the institutes include Biodiversity Conservation, Germplasm supply and distribution, Genetics, Tree breeding, Biotechnology, Domestication and Conservation of Forestry Genetic Resources.

iii. Institute for Commercial Forestry Research, South Africa

This institute was founded in the year 1947 as the Wattle Research Institute and upgraded as Institute for Commercial Forestry Research. The major tree improvement activities of the institute incorporated Eucalyptus and Wattle to develop genetically superior plant extracts coupled with varieties for pest and disease tolerance. Establishment of seed orchard to supply quality seed is also one of the mandated objectives of the institute.

iv. Forestry and Forest products Research Center, Durban

This center is established at Durban and incorporates two different centers as indicated below:

i. Forest Tree Breeding Center (FTBC)

ii. Forest Bio Research Center (FBRC)

The major activities of the center include development of genetically superior varieties, accelerating tree breeding process, collection, conservation and preservation of genetic materials followed by establishing research cooperation with other developing countries.

v. National Forestry Resources Research Institute (NaFORRI)

The major activities of the institute include tree improvement leading to selection of most desirable species. Intensive breeding programme and testing the improved progenies and establishment of seed orchard for seed production and development of mother clonal garden are the major activities to produce the quality seedlings and clonal materials. This besides, the application of biotechnological tools to quantify the genetic variation and to speed up the traditional tree breeding programme through markers assisted selection formed a major research activity. In addition, development of breeding strategies, establishment of gene banks and studies related to breeding biology are the major activities of the institute.

vi. Forestry and Agricultural Biotechnology Research Institute, University of Pretoria

This institute was established under the agies of the University of Pretoria with a main focus on academic research programme in forest molecular genetics. The major focus of research programme includes Gene discovery, functional genetic research recovery like transcriptomics, metabolomics, wood microchemistry, plant transformation etc. This besides, allele discovery and development of molecular breeding tool for genetic improvement of target forestry species are also included as research priority areas. It was established in the year 1997 and other research areas include the following

- Agriculture and forest biotechnology

- Forest entomology and forest pathology
- Microbial bioinformatics and cytogenetics
- Mycology and fungal biodiversity
- Biological control of fungal and insect pests

vii. Forest Research Institute Malaysia (FRIM)

This institute was established in the year 1929 and acts as a leading institute of tropical forestry research. The institute is involved in generating scientific knowledge for understanding management, conservation and use of forest resources. The study of biodiversity forms a major parts of activity; development of technologies to cater to the forest based breeding technologies is also carried out by this institute. Above all the institute is involved in establishing linkage with global research organizations.

viii. The Federal Forest Research Centre

It was established in the year 1874 in Austria and mandated to protect and conserve the forest genetic resources. It is renamed as the Federal Research and Training Centre for Forests, Natural Hazards and Landscape in the year 1962.

ix. Forestry Research Institute, Sweden

This research institute was established in Uppasala, Sweden and mandatory to provide the industries with knowledge, product, product services through cost effective and sustainable forestry activities. The major research efforts are concentrated on tree breeding with a focus on development of best planting material with better growth, resistance and wood quality. The major species of interest primarily are Scots pine, Norway spruce, Iodgepole pine and birch, hybrid ash, Douglas fir, larch and Sitka spruce.

x. Biodiversity International

It is a CGIAR institution established in 1974 as International Bureau of Plant Genetic Resources (IBPGR). This institute was renamed as International Plant Genetic Resources Institute (IPGRI) in October 1993 and became independent in 1994 as one of the centers of CGIAR. The major activity of the institute is safeguarding and use of biological diversity to meet the challenges through adaptation to climate change and increased sustainable production. This CGIAR institution take into account all the diversity of plants and animals and facilitates through breeding and delivering other benefits like better nutrition, greater sustainability and ecosystem services. The institute is located in Rome, Italy.

xi. Center for International Forestry Research (CIFOR)

It is the Center for International Forestry Research with the headquarters in Bogor, Indonesia. It has 3 regional offices and several project offices. This institution is one of the 15 centers within the CGIAR institutions and involved in multidisciplinary approach towards improvement and management of tropical forest and addresses the issues of the people. It was establish during 5th March 1993 as Non-profit organization under the agies of CGIAR with a focus on Forest, trees and agroforestry. This institution is incorporated with the following 5 major research programme

- Component 1 - Small holder production system on market
- Component 2 - Management and conservation of forestry and tree resources
- Component 3 - Environmental services and landscape management
- Component 4 - Climate change adaptation and mitigation
- Component 5 - Impacts of trade and investment on forest and people

2. NATIONAL INSTITUTES

i. Forest College and Research Institute, TNAU

The institute was established as a department in the year 1973, upgraded in to Forestry Research Station in the year 1976 followed by its upgradation as Forest College and Research Institute in the year 1990. The institute is mandated with the tree breeding programmes to develop varieties amenable for varied industrial utility like timber, plywood, pulpwood, match splints, dendro power, bio fuel, non timber forest products etc.,.

The breeding activities include selection and hybridization followed by systematic evaluation through progeny and clonal test. Development of protocol for mini clonal technology for wide range of species and establishment of seed orchard coupled with development of seed production technologies are the major areas of tree breeding research reorganized by this institute. The species like Teak, Red sanders, Sandal, Albizia, Acacia, Bamboo, Casuarina, Dalbergia, Eucalyptus, Melia, Mulberry, Gmelina, Kadam, Sterculia, Toona, Acrocarpus, Mahogany, Silver oak, Bio fuel species like Jatropha, Mahua, Pungam and NTFP species like Silk Cotton, Bixa, Tamarind, Medicinal plants etc., are incorporated in the breeding and improvement programme. The institute has developed so far eight High Yielding Short Rotation Variety representing 6 different species.

ii. Indian Council of Forestry Research and Education (ICFRE)

ICFRE is a government organization and established 9 research institutes across the country and involved in tree breeding and improvement programme. The headquarter is at Dehradun and in the headquarter tree breeding is a major theme of research and prioritized species like Poplars, Eucalyptus, Sisham, Bamboos and few temperate species. Some of the ICFRE institutes involved in major tree breeding programme are furnished below

a. Institute of Forest Genetics and Tree Breeding (IFGTB), Coimbatore

The IFGTB was established in the year 1988 as a National Institute under the ICFRE. The institute has mandated tree breeding as the core area of research and involved in varietal developmental programme of wide range of species like Casuarina, Eucalyptus, Teak, Acacia, Gmelina etc.,. The institute has also developed high yielding short rotation clones in Casuarina and Eucalyptus and released them for used by different stake holders. This besides, the institute also involved in clonal technology development, forest biotechnology and forest genetic resources conservation through *in-situ* and *ex-situ* conservation approaches.

b. The Institute of Forest Biodiversity, Hyderabad

It is one of the ICFRE institutes established in the year 2012 and mandated to undertake and promote Scientific Research for conservation and management of biodiversity including marine and coastal resources. Assessment of biodiversity, documentation, conservation, sustainable utilization and invasive species are the major areas of breeding research carried out by this institute.

iii. Kerala Forest Research Institute (KFRI), Thrissur, Kerala

The institute was established in the year 1975 with the mandated research on tropical forest research. Tree breeding improvements are the major activities of the institute which leads to genetic improvement of species like Teak, Eucalyptus, Melia, Dalbergia, Mahogany, Ailanthus, Terminalia etc. Studies on reproductive biology, diversity assessment through biotechnological tools, karyotype analysis, photochemical analysis etc., are also the major areas of tree breeding research operated by this institute.

iv. National Bureau of Plant Genetic Resources (NBPGR), New Delhi

The National Bureau of Plant Genetic Resources has its inception as a division of Plant Introduction under the agies of Indian Agricultural Research Institute in 1961. Before that the division was under the botany with plant introduction and exploration organization. Based on the recommendation of high level committee constituted by GOI, an independent institute viz., National Bureau of Plant introduction was established in the year 1976. The institute name was rechristened into National Bureau of Plant Genetic Resources during January 1997 with the head quarters in New Delhi. The NBPGR has 10 regional stations representing almost all parts of India. The major activities of NBPGR include the following

- Exploration, collection, characterization and maintenance of various crop varieties
- Improvement of various crop plants
- Diversification and development of agriculture

- Introduction and exchange of genetic resources both within and outside the country.
- Conservation of genetic resources both by traditional and also with modern biotechnology tools

v. National Institute of Plant Genomic Research (NIPGR)

This institute is established as National Center for Plant Genome Research under the agies of DBT during 30th November 1997. Then the institute is upgraded as National Institute of Plant Genomic Research with the mandate to emerge as one of the important resource institute for material, knowledge and technologies in the areas of functional, structural, evolutionary and applied genomics of plants including crop plants.

The major areas of research include computational biology, genome analysis and molecular mapping, molecular mechanism of abiotic stress response, Nutritional genomics, plant development and architecture, plant immunity, molecular breeding, transgenic for crop improvement and other emerging areas of plant genomics.

vi. National Research Centre on Plant Biotechnology (Centre for Plant Biotechnology)

It was established in the year 2000 at Hisar, Haryana under the agies of Agricultural research. The institute is then taken by Government of Haryana in the year 2005 and renamed as Centre for plant biotechnology in the year 2007. The major areas of research include genetic diversity analysis, propagation through tissue culture technology, training and mass propagation of planting material of newly developed fruits, crop varieties and rare medicinal plants, horticulture, forest, ornament and other plant species.

SUMMARY

The success of any plantation development programme depends primarily on the genetic resources used in plantation development activities. The productivity and profitability of several plantations are widely questioned due to poor survival, low productivity and existence of variable yield estimations. This has demanded deployment of organized tree breeding programme which has successfully proved significant yield improvement, wood quality manipulation and resistance towards biotic and abiotic factors. The organized tree breeding and improvement with focused objective is the primary aim of several institutes, organizations and wood based industries. The institutes and organizations with well defined organizational structures, focused objectives and efficient research management have led to significant research outputs in the form of varietal development leading to higher productivity and profitability. Several institutions distributed across the world have been collected and presented in this chapter as a ready reckoner for the tree breeders to learn the-state-of-the-art- developments in the tree breeding across the globe. It will also help the tree breeders to establish suitable linkages for capacity building, germplasm exchange and to introduce new species and varieties suitable for varied edapho-climatic conditions.

CONVENTIONS AND PROTOCOLS

19

INTRODUCTION

The biological diversity which include genetic differences between, within the species and between the species in a variety of ecosystem. The term biodiversity deals with wide variety of life forms on the earth which include plants, animals and micro organisms as well as the interacting ecosystem. The interaction between biological components and the ecosystem provides wide range of ecosystem services such as clean air, water, food, medicine, fuel, fodder, fibre, timber and other direct and indirect products. Such a multifarious biodiversity resources have been lost at unprecedented rates due to anthropogenic pressure, industrialization, urbanization, intensification of agriculture, deforestation, encroachment and climate change related risks and uncertainties. To control these issues and to conserve the biological diversity, the United Nation Environment Programme established an adhoc working group on biodiversity for the harmonization of existing convention related to biological diversity. The experts of adhoc group agreed in principle on the need to elaborate an international binding instrument on biological diversity. Accordingly during May 1989 another adhoc working group on biological diversity was established to prepare an international legal instrument for their conservation and sustainable use of biodiversity. It has also been suggested to take care of the issues related to sharing the cause and benefit between the countries and the means to support innovations of the local people.

This adhoc group has been upgraded into Intergovernmental Negotiation Committee (INC) during February, 1991 and held series of working sessions including negotiation meetings which culminated and the final and agreed text of Conventional on Biological Diversity was developed at the Nairobi final act of the conference for adoption of the agreed text of the convention. Then the convention was opened for signature at the United National Conference on Environment and Development (UNCED) in Rio De Janeiro, Brazil in June 1992 which has entered into force on 29th December 1993.

Based on this convention on biological diversity, several other conventions and protocols have also been emanated for the conservation, sustainable use of biodiversity, benefit sharing mechanism, sharing of resources and technologies and the Biosafety of products emanated from modern biotechnology. Hence the current chapter on conventions and protocols have been designed and presented with 4 conventions and 2 protocols which are direct relevant to the current book on Tree breeding and improvement.

DEFINITION OF TERMINOLOGIES

Convention	:	A Convention is a meeting or gathering to formulate or deliberate on a generally accepted principle, a framework in which the parties decide the basic guidelines.
Conference	:	A Conference is consulting together formally; a type of negotiations, there is only broad theme.
Protocol	:	A Protocol to the convention is an agreement that diplomatic negotiators formulate and sign as the basis for a final convention where the parties set specific aims or legal obligations.
Conference of Parties	:	A conference of the parties (COP) is the governing body of an international convention
Treaty	:	A treaty is a formal written agreement entered into by actors in international law, namely sovereign states and international organizations. A treaty may also be known as an international agreement, protocol,

		covenant, convention, pact, or exchange of letters, among other terms.
Biosafety	:	Biosafety is a term used to describe efforts to reduce and eliminate the potential risks resulting from biotechnology and its products.
Living Modified Organism	:	A Living Modified Organism (LMO) is defined in the Cartagena Protocol on Biosafety as any living organism that possesses a novel combination of genetic material obtained through the use of modern biotechnology.
Biodiversity	:	Biodiversity is the variety and variability of life on Earth. Biodiversity is typically a measure of variation at the genetic, species, and ecosystem levels.

1. CONVENTIONS

i. CONVENTION ON BIOLOGICAL DIVERSITY (CBD)

The CBD is an international legally binding treaty which was opened for signature at the Earth Submit held in Rio de Janeiro on 5th June 1992. The treaty was entered into force from 29th December 1993 onwards and as on date there are 193 parties signed the treaty.

This Convention on Biological Diversity comprises 42 articles and three annexures. The articles 1 of the CBD deals with the objectives of the CBD which primarily concentrated on following 3 main goals

- Conservation of Biological Diversity
- Sustainable use of Biological Diversity and its component
- Fair and equitable sharing the benefit arising from the utilization of genetic resources including appropriate access to genetic resources and transfer of relevant technologies.

SALIENT FEATURES OF CONVENTION ON BIOLOGICAL DIVERSITY

Sustainable use of Biological Diversity

- The article 6 of CBD provides the general measures for conservation and sustainable use of biological diversity and directed all contracting parties to develop national strategies plans and programmes for the conservation and sustainable use of biological diversity.

In-situ and _ex-situ_ conservation

- Article 8 and 10 deals with the conservation of biological resources both through *in-situ* and *ex-situ* conservation. The contracting parties should establish guidelines for selection, establishment and management of protected areas towards *in-situ* conservation of biological diversity.
- The contracting parties should establish and implement a plan to rehabilitate and restore degraded ecosystem and promote the recovery of threatened species.
- The contracting parties should established strategies to prevent to introduction, control and eradicate the invasive alien species which threatened the ecosystem, habitat and species.
- Similarly the contracting parties should adopt measures for *ex-situ* conservation of biological diversity by establishing necessary infrastructure facilities, adopting measures for recovery and rehabilitation and establish co-operation in providing financial and support services towards *ex-situ* conservation of biological resources.

Incentive Measures

- The article 11 of CBD direct that each contracting parties should adopt economically and socially sound measures that acts as incentives for the conservation and sustainable use of biological diversity.

Research and Training

- The article 12 of CBD direct that the contracting parties should establish and maintain programmes for Scientific and technical education to train the people towards identification, conservation and sustainable use of biological diversity.
- It also indicates that the country should promote and encourage research towards conservation and sustainable use of biodiversity.
- The article 12 also directs the developed countries to support the developing countries in terms of scientific and technical

education coupled with research and establish co-operation among member countries towards creating scientific advances for conservation and sustainable use of biological uses.

Access to Genetic Resources

- The article 15 of CBD indicates that each contracting party shall create the condition to facilitate to access the genetic resources based on mutually agreed terms subject to the provision of the article.

Access to transfer technology

- The article 16 of the CBD directs the contracting parties towards extending transfer of technologies including biotechnology that are relevant to conservation and sustainable use of biological diversity and will not cause damage to environment.
- In case of transfer of technologies to developing countries, the most favorable terms and conditions including concessional and preferential terms should be extended.
- The article also directing the contracting parties to take necessary legislative, administrative and policy measures in order to protect the patents and other Intellectual Property Rights generated through the use of biological resources and transfer of technology.

Financial Resources

- • The article 20 of CBD directs the parties representing in developed countries who shall provide new and additional financial resources to the parties of developing countries to meet the provision and agreed terms and conditions of CBD.

ii. RAMSAR CONVENTION

It is the convention on wetlands adopted on 2nd February, 1971 as an intergovernmental treaty. This was adopted in the Iranian city of Ramsar on the southern shore of Caspian Sea.

Though the name of the convention is Convention on Wetlands but it has been popularly pronounced as Ramsar Convention.

Scope and importance of the convention

The convention on wetlands of international importance especially for water fowl habitat reflects the original emphasis upon the conservation and wise use of wetlands primarily as a habitat for water birds. However over the years, the convention has broadened its scope and covered all aspects of wetland conventions.

The convention came into force in 1975 and as of 2016, 169 countries signed the treaty and become member of the contracting parties.

SALIENT FEATURES OF THE CONVENTION

- It is estimated that globally 2220 wetlands are identified as Ramsar cites and covers an area of 214 million ha.
- The article 2 of the convention listed the wetlands of international importance and directed the contracting parties to promote its conservation and continue to designate suitable wetlands within its tertiary.
- The article 3 of the convention indicated that the contracting parties are to include the wetland conservation in all their national planning process. The contracting parties should also formulate and implement the plan and strategy towards the wise use of wetlands in their tertiary.

Ramsar cites in India

- In India the convention came into force from 1st February 1982.
- India is housed with 27 cites designated as wetlands of international importance and covers an area of 689131 ha.

S.No	Name	Place	Designated	Area (km^2)
1.	Chilika Lake	Odisha	1 October 1981	1165.00
2.	Keoladeo National Park	Rajasthan	1 October 1981	28.73
3.	Harike Wetland	Punjab	23 March 1990	41.00
4.	Loktak Lake	Manipur	23 March 1990	266.00
5.	Sambhar Lake	Rajasthan	23 March 1990	240.00

S.No	Name	Place	Designated	Area (km²)
6.	Wular Lake	Jammu and Kashmir	23 March 1990	189.00
7.	Kanjli Wetland	Punjab	22 January 2002	1.83
8.	Ropar Wetland	Punjab	22 January 2002	13.65
9.	Ashtamudi Wetland	Kerala	19 August 2002	614.00
10.	Bhitarkanika Mangroves	Odisha	19 August 2002	650.00
11.	Bhoj Wetland	Madhya Pradesh	19 August 2002	32.00
12.	Deepor Beel	Assam	19 August 2002	40.00
13.	East Kolkata Wetlands	West Bengal	19 August 2002	125.00
14.	Kolleru Lake	Andhra Pradesh	19 August 2002	901.00
15.	Point Calimere Wildlife and Bird Sanctuary	Tamil Nadu	19 August 2002	385.00
16.	Pong Dam Lake	Himachal Pradesh	19 August 2002	156.62
17.	Sasthamkotta Lake	Kerala	19 August 2002	3.73
18.	Tsomoriri	Jammu and Kashmir	19 August 2002	120.00
19.	Vembanad-Kol Wetland	Kerala	19 August 2002	1512.50
20.	Chandra Taal	Himachal Pradesh	8 November 2005	0.49
21.	Hokera Wetland	Jammu and Kashmir	8 November 2005	13.75
22.	Renuka Lake	Himachal Pradesh	8 November 2005	0.20
23.	Rudrasagar Lake	Tripura	8 November 2005	2.40
24.	Surinsar-Mansar Lakes	Jammu and Kashmir	8 November 2005	3.50

S.No	Name	Place	Designated	Area (km²)
25.	Upper Ganga River (Brijghat to Narora Stretch)	Uttar Pradesh	8 November 2005	265.90
26.	Nalsarovar Bird Sanctuary	Gujarat	24 September 2012	123.00
27.	Sundarban Wetland	West Bengal	1 February 2019	4230.00

(*Source:* https://rsis.ramsar.org/sites)

iii. CONVENTION ON INTERNATIONAL TRADE IN ENDANGERED SPECIES OF WILD FAUNA AND FLORA (CITES)

The Convention on International Trade in Endangered Species of Wild Fauna and Flora (CITES) is otherwise pronounced as Washington Convention. The convention was signed on 3rd march 1973 and became into force from July 1995 onwards.

Scope of convention

- This convention provides an international legal framework for the regulation of trade of plants and animal species which are commercially exploited for international trade.
- The CITES listed three appendices incorporating wide range of species of flora and fauna as indicated below.
    - **Appendix I:** It includes plants species threatened with extinction for which international trade is straightly regulated. Nearly 300 plant species are included in the appendix I
    - **Appendix II:** It includes species that are not threatened with extinct at present but may become so if unregulated trade continues. There are over 28,000 species listed in appendix II which include entire orchid and cactus family.
    - **Appendix III:** It list species that are threatened locally with extinction through commercial exploitation and therefore these species are subject to trade control within certain nations. Only few and handful of species are alone listed.

- This convention operates through the issue and control of import and export permit for a wide range of species listed in the appendices.
- The CITES also regulates trade of plant species which can withstand and tolerate the current rate of exploitation
- CITES prevents trade of those species that are threatened with extinction.

As for as trees are concerned there are about 80 tree species included in the CITES appendices.

Countries signed in the treaty

As of now, 183 countries signed the treaty and involved in implementation of CITES guidelines and strategies. India signed the treaty and became a party of the convention in the year 1976.

SALIENT FEATURES OF CITES

- The CITES conventions incorporated 25 article with three appendices for the list of species and the 4th appendices for model export permit.
- The article II of CITES defined fundamental principles and directed the contracting parties that the parties shall not allow trade in specimens of species included in appendix I, II and III except in accordance with the provision of present conventions.
- The article III indicates the regulation of trade in specimen of species included in appendix I. The export of specimen included in appendix I require prior grant and presentation on export permit granted by scientific authority of the state of export or the management authority of the state of export. Similarly the import of any specimen of appendix I require prior grant and presentation of import permit issued by scientific authority of the state of import or management authority of the state of import.
- The article IV of CITES convention deals with the regulation of trade in specimens of species included in appendix 2. Similar import and export permit issued by concerned authority of the import and export country is essential.

- The article V of the CITES convention deals with the regulation of trade in specimen of species included in appendix III.
- The exports of specimen listed in appendix III require prior grant and presentation of an export permit granted by the management authority of the state of export.
- The import of any specimen of a species included in appendix III shall require, except in circumstances to which paragraph 4 of this article applies, the prior presentation of a certificate of origin where the import is from a state which has included that species in appendix III, an export permit.
- The article VI deals with various permits and certificates related to import and export of specimens of species listed in various appendices.
- The article VII of the CITES convention deals with measures taken to by the parties particularly the party should maintain records of trade in specimens of species included in appendix I, II and III.
- The article X deals with the trade of specimen to the country where it is a part of the convention.
- The article XVII deals with resolution of disputes which may arise between parties with respect to the applications of provision of the present convention and suggest that mutual negotiation between the parties to resolve the issue.

iv. CONVENTION ON MIGRATORY SPECIES OF WILD ANIMALS (CMS)

The Convention on Migratory Species on wild animals is also referred as Bonn convention and signed in 1979 and came into force 1983 onwards.

SALIENT FEATURES OF THE CONVENTION

- This conventions aims to conserve terrestrial, aquatic and avian migratory species thought the range.
- The convention facilitates the adoption of strict protection measures for endangered migratory species, multilateral agreements and co-operative research activities towards conservation and management of migratory species.

- This convention / treaty have been signed by 129 countries as of now including India and became a party of the convention.
- This convention establishes collaboration and co-operation with several organizations and institutions distributed across the world towards management and conservation of migratory species.
- This convention has incorporated the following two appendices
  - o **Appendix I:** It lists migratory species that are classified as endangered which demand urgent international co-operation to resolve the issues.
  - o **Appendix II:** It lists other species that require international attention towards conservation.
- In India the marine turtle, Siberian crane, dudang are the part of conservation measures through this convention.

2. PROTOCOLS

There are two major two protocols relevant to tree breeding and improvement. They are

i. CARTAGENA PROTOCOL

This protocol is based on the recommendation of the Convention on Biological Diversity to address the issues of modern biotechnology to protect human health and the environment and possible adverse effect of the products through modern biotechnology. The CBD recognizes two major aspects of modern biotechnology as indicated below

- The CBD facilitates for the access to a transfer of technology including biotechnologies that are relevant to conservation and sustainable use of biodiversity as per the provision of article 16 and 19 of the Convention on Biological Diversity.
- The article 8 (g) and the article 19 (Paragraph 3) seek to ensure the development of appropriate procedures to enhance the safety of biotechnology in relation to human health

Considering these two aspects into account, the Convention on Biological Diversity directs the parties for the development of an international legally binding instrument to address the issue of

Biosafety. Accordingly the Conference of Parties at its second meeting held in November 1995 established an adhoc working group on Biosafety to develop draft protocol specifically on Biosafety with a focus on transboundary movement of any living modified organism resulting from modern biotechnology which may have adverse effects.

After several round of negotiation and discussion, the Biosafety protocol commonly pronounced as Cartagena protocol on Biosafety was finalized. This protocol was adopted in Montreal on 29th January, 2008 at an extraordinary meeting of the conference of parties.

SALIENT FEATURES OF THE PROTOCOL

- The protocol incorporates 40 articles and 3 annexures.

Biosafety Protocol Article – 2

- Each party is required to take necessary and appropriate legal, administrative and other measures to implement its obligations under the Protocol and the Parties are Required to ensure that the development, handling, transport, use, transfer and release of Any living modified organisms are undertaken in a manner that prevents or reduces The risks to biological diversity, taking also into account risks to human health

Biosafety Protocol Articles 7, 8

- Exporting countries shipping living modified organisms (GMOs) for intentional introduction into the environment will have to give prior notification of the importing country that is a party to the protocol under the Advance Informed Agreement (AIA) procedure and provide sufficient information to enable them to make informed decisions.

Biosafety Protocol Article 18

- Shipments of GMOs intended for intentional introduction to the environment will have to be identified in accompanying documentation as GMOs with specification of the identity and characteristics and with a declaration that " the movement is in conformity with the requirements of the Protocol"

Biosafety Protocol Articles 11, 18, Annex II

- Shipments of GMO commodities intended for direct use for food, feed or processing will, in the interim, have to be identified in accompanying documentation that they "may contain" GMOs and are "not intended for intentional introduction into the environment"
- Any party that approves for domestic use and marketing GMOs intended for direct use as food, feed or processing that may be exported will be required to communicate this decision and details about the GMO to the world community via an electronic, web-accessible database: the Biosafety Clearing-House.

Advanced Informed Agreement (AIA)

i. Obtain the importing country's consent

ii. 270 days – shipment period

iii. Importing country – make necessary preventive measures due to GMOs

iv. Importing country can any time review and change the decision of import If there is lack of scientific certainty regarding the extend of any potential adverse effect

Labeling Requirements

GMOs intended for direct introduction into the environment (deliberate release)

The Protocol required exporter to ensure that all consignments of GMOs that are intended for direct introduction into the environment are clearly labeled

Label

- Identification of the consignment as 'Living Modified Organisms'
- Specific identity and relevant traits and/or characteristics
- Any requirements for safe handling, storage, transport and use
- The contact point for further information, and as appropriate, the name and address of the importer and exporter

- A declaration that the movement is in conformity with the requirements of the Protocol (i.e that the importing country has given permission for the consignment to enter its territory)

GMOs Intended for Contained Use

- The labels for GMOs intended for contained use do not have to include the Specific identity, relevant traits and/or characteristics of the GMO. They must, however Include:
- Identification of the consignment as 'living modified organisms'
- Any requirements for safe handling, storage, transport and use
- A contact point for further information (Including the name and address of the individual and institution to whom the GMO is being sent)

BIOSAFETY REGULATIONS IN INDIA

India has severe regulation procedures for Biosafety aspects of modern biotechnological products through its ratification on Cartagena protocol on January 23, 2003. To ensure Biosafety Govt. of India implemented the following legal provisions

i. Environmental Protection Act, 1986

- MoEF, Govt. of India notified rules and policies to handle GMOs under environmental Protection Act 1986

ii. Under Rule 1989

- No person shall handle GMO's without approval of GEAC
- GMOs for research allowed under notification of EPA 1986
- Education and experiments on GMOs look after by IBSC
- Unintentional / deliberate release of GMOs not allowed
- GEAC supervises, rules and guidelines implementation
- If orders not followed. GEAC take suitable action
- All approvals are for 4 years (renewable for 2 years)
- GEAC have powers to revoke approvals

iii. Seed Policy 2002

- GM crops tested for environment and Biosafety before release as per EPA, 1986
- After release, its seeds registered and marketed as per Seed Act
- Seed Policy (2002) contains separate section for GM crops

iv. Recombinant DNA Guidelines, 1990

- DBT formulated rDNA guidelines in 1990 for research
- Revised in 1994, to cover R & D on GMOs
- Levels of risk and classification of organism defined by guidelines
- Guidelines employ physical and biological containment
- And also principles of good laboratory practices (GLP)
- Safety criteria defined for large scale practice (>20 acres)
- It specify, containment to release into the environment
- Independent review conducted on case to case basis.

v. Transgenic Plant Guideline, 1998

- Revised guidelines for research in transgenic plants
- Special MEC to monitor the impact on environment (Monitoring and evaluation committee)
- Also collects and reviews the information on transgenic plants.

Competent Authorities

- **IBSC** – Institutional Biosafety Committee
- **GEAC** – Genetic Engineering Approval Committee
- **DLC** – District Level Committee
- **SBCC** – State Biotechnology Co-ordination Committee
- **RCGM** – Review Committee on Genetic Manipulation
- **RDAC** – Recombinant DNA Advisory Committee

Risk Assessment

1. **Initial Risk Assessment**

   Made by investigator based on risk group (RG)

   4 RGs based on their effect on healthy human

- RG_1 → **No disease** on healthy human
- RG_2 → **Non serious disease** on healthy human, therapy available
- RG_3 → **Serious disease** on human, therapy may be available
- RG_4 → **Serious disease** of human, therapy may not available

2. **Comprehensive Risk Assessment**
   - Include features of organisms
   - Virulence, pathogenicity, infectious dose, route of spread, communicability, treatment availability etc.
   - Any strain more hazardous than parent handled at higher containment level
   - Exotic pathogens restricted

Risk Assessment for Planned Release

- Biological characters of Recipient. O
- Biological characters of Donor. O
- Characters of vector
- Characters of promoter
- Genetic characters of DNA insert
- Nature of Marker gene
- Proximity of related & wild species
- Chances of inter pollination and their survival

Risk Assessment for Products

- Whether foreign DNA encode harmful substances
- Safety of protein encoded
- Proof for acceptable level of known plant toxicants
- Food crop that is modified

Risk Regulation

- Biosafety during laboratory research to minimize the risk procedures, practices & requirements etc., developed

- Procedure to import GMO for research different (Ghosh *et al.*, 2000).

1. **Physical containment**
    - Special laboratory design and equipment
    - Special procedure to restrict accidental release of organism
    - Organism categorized as BL_1, BL_2, BL_3 & BL_4
    - As the BL increases, laboratory design & procedures become more rigid
    - Emergency plan in case of accidental containment
    - Everyone should familiar with hazards and emergency plan
    - Vaccinate all workers , if known pathogen involved.

ii. NAGOYA PROTOCOL

This protocol forms a part of convention on biological diversity which was designed as protocol at the 10th Conference of Parties (COP) held in Nagoya, Japan during October, 2010. The primary objective of this protocol is to establish access to genetic resources and to adopt fair and equitable sharing of benefits arising out of the utilization of biological resources.

SALIENT FEATURES OF THE PROTOCOL

- It is an international agreement and treaty deals with benefit sharing through the use of genetic resources in a fair and equitable way there by contributes towards conservation and sustainable utilization of biological diversity.
- This protocol builds on the access and the benefit sharing provisions of the CBD by establishing greater legal certainty.
- It also creates transparency among the provider and users of the genetic resources.
- The protocol also establishes more predictable conditions for access to genetic resources and help to ensure sharing of benefits particularly when the genetic resources leave the country of genetic resources origin and supplier.

- This protocol also ensures protection of traditional knowledge associated with genetic resources that comes from indigenous and local communities.
- The protocol also establishes provision for the use of plant animals and or microorganism for use in wide range of purposes ranging from basic research to development of products and ensures provision of benefit sharing, royalty payment and other IPR issues.
- The protocol also facilitates and ensures joint ownership of intellectual property rights particularly when the user and the provider of genetic resources seek joint ownership of IPR for patented products based on the resources used.
- The protocol also facilitates research exchanges between the provider and user countries.
- The protocol also establishes directives for the providers and user countries. When an institution seeks access to genetic resources from other country, it should obtain the prior informed concerned of the country in which the resource is located. As per the provision of protocol, the importing institution may also negotiate and agree on the terms and conditions of the access and use of this resource.
- Similarly the providing country should give fair and non-orbitry rules and procedures for access to their genetic resources.

SUMMARY

This chapter has incorporated 4 conventions and 2 protocols along with the history, development and salient features. The Convention on Biological Diversity which is pioneer for all other conventions and protocols predominantly focused on conservation and sustainable use of biological diversity. This convention has led to creation of two protocols viz., Cartagena protocol on Biosafety and Nagoya protocols for fair and equitable share of benefit of the product arising out of utilization of biological resources. The complete features of both Biosafety and benefit sharing aspects has been furnished which will act as a master piece of information for the students, researchers, policy makers and bureaucrats. The incorporation of other conventions like Ramsar convention, convention on Migratory birds and Convention on International

Trade in Endangered Species of Wild Fauna and Flora (CITES) in this chapter will help the readers to understand the existing legal and policy provisions towards conservation and sustainable utilization of genetic resources in general and migratory birds, wet land ecosystem in particular. The information furnished on CITES will be of more useful to the readers to know the status of handling of specimens listed in annexure I, II and III of CITES for trade through international market mechanism. This chapter will have profound utility for all type of readers not only for academic utility but also for research purposes leading towards conservation and sustainable utilization of biological resources.

BIBLIOGRAPHY

Anil Kumar and Priti Maheshwari, 2003, Textbook of biotechnology, Agrotech publishing Academy, Udaipur. P:480 .

Arcade, F., Anselin, P., Lesage, F.R.M.C., Paques, L.E. and Prat, D. 2000. Application of AFLP, RAPD and ISSR markers to genetic mapping of European and Japanese larch. *Theor. Appl. Genet.* 100: 299-307.

Bhabendra Kumar, Patnaik, Tarani Charan Kara, Soumendra Nath Ghosh and Abhaya Kumar Dalai, 2012, Text Book of Biotechnology, Tata McGraw Hill Publishing Company Limited New Delhi.

Bhojwani, S.S., 1980. Micropropagation method for a hybrid willow (Salix matsudana× alba NZ-1002). *New Zealand Journal of Botany, 18*(2), pp.209-214.

BIO.FSU.EDU. 2013. Chapter 4: Pollen vectors and pollination syndromes. Retrieved Nov. 24, 2013 from http://bio.fsu.edu/~winn/3402L/WinnCH4.html.

Bracelpa, 1999 The annual report of the Brazilian Pulp and Paper Association. Sao Paulo, Brazil.

Brondania, R.P.C., Brondani, R. and Grattapagalia, T.D. 1998. Development, characterization and mapping of microsatellite markers in *Eucalyptus grandis* and *E. Urophylla. Theor.Appl. Genet.*97: 816-827.

Campinhos E and Ikemori YK 1983 Mass Production of *Eucalyptus* spp. by rooting cuttings. Silvicultura 8 (32):770-775

Cervera, M.T., Gusmão, J., Steenackers, M., Van Gysel, A., Van Montagu, M. and Boerjan, W., 1996. Application of AFLP™-based molecular markers to breeding of Populus spp. *Plant Growth Regulation, 20*(1), pp.47-52.

Chauhan, S., Chauhan, S.V.S. and Galetto, L., 2017. Floral and pollination biology, breeding system and nectar traits of Callistemon citrinus (Myrtaceae) cultivated in India. *South African Journal of Botany, 111*, pp.319-325.

Cláudia Inês da Silva and Helena Maura Torezan Silingardi. Reproductive Biology Of Tropical Plants, International Commission on Tropical Biology and Natural Resources.

Darbeshwar Roy, 2012, Plant breeding - A biometric Approach, Narosa Publishing House Pvt Ltd., Mumbai

Dhamu KP, Ramamoorthy K. Statistical methods. Agrobios (India), Jodhpur. 2007, 236.

Dhayananthan, M, 2011. Studies on invitro propagation, cytogenetics and molecular analysis of Jatropha genetic resources. *M.Sc Thesis, Tamil Nadu Agricultural University, Coimbatore, India.*

Dhixya Deve, K. R. 2014. Studies on Genetic Variation, Wood and Molecular Characterization of *Dalbergia sissoo* Roxb. Genetic Resources. *Ph.D Thesis, Tamil Nadu Agricultural University, Coimbatore, India.*

Dieters, M. and Brawner, J., 2007. Productivity of Pinus elliottii, P. caribaea and their F 1 and F 2 hybrids to 15 years in Queensland, Australia. *Annals of forest science, 64*(7), pp.691-698.

Dorofeyev, V.F and A.A. Filatenko. 1992. Preface xix-xxxi. N.I. Vavilov. Origin and geography of cultivated plants. In: V.F. Dorofeyev (ed.), translated by D. Love. University of Cambridge University Press, Great Britain.

Feng, S.P., Li, W.G., Wang, H.S.H.J.Y. and Wu, Y.T. 2009. Development, characterization and cross-species/ genera transferability of EST-SSR markers for rubber tree. *Mol. Breed.* 23: 85-97.

Gamborg, O.L., Miller, R.A. and Ojima, K. (1968) Plant Cell Cultures 1. Nutrient Requirements of Suspension Cultures of Soybean Root Cells. Experimental Cell Research, 50, 151-158.

Geetha S, Jebaraj. S and Pandiyarajan P. 2007. Agricultural Biotechnology. Agrobios (India), Jodhpur. P: 275.

Geetha, S., Jebaraj, S. and Pandiyarajan, P., 2008. Agricultural biotechnology. Agrobios (India).P:306

George E.F.., 2008, Plant Tissue Culture Procedure - Background, Ed: George, E.F., Hall, M.A. and De Klerk, G.J., 2008. Plant tissue culture procedure-background. In Plant propagation by tissue culture (pp. 1-28). Springer, Dordrecht.

Ghazoul, J. and McLeish, M., 2001. Reproductive ecology of tropical forest trees in logged and fragmented habitats in Thailand and Costa Rica. *Plant Ecology, 153*(1-2), pp.335-345.

Gopalakrishnan, K.K. and Thomas, T.D., 2014. Reproductive biology of *Pittosporum dasycaulon* Miq., (Family *Pittosporaceae*) a rare medicinal tree endemic to Western Ghats. *Botanical studies, 55*(1), p.15.

Gupta, S.K. ed., 2000. Plant breeding: Theory and techniques. Agrobios.

Haines, R., 1994. Biotechnology in forest tree improvement with special reference to developing countries. p: 230.

Hummer, K.E. and J.F. Hancock. 2015. Vavilovian centers of Plant diversity: Implications and Impacts. Hort science, Vol. 50 (6) : 780-783.

Hussain, A., Qarshi, I.A., Nazir, H. and Ullah, I., 2012. Plant tissue culture: current status and opportunities. In Recent advances in plant in vitro culture. IntechOpen.

James, R.R., Croft, B.A. and Strauss, S.H., 1999. Susceptibility of the cottonwood leaf beetle (Coleoptera: Chrysomelidae) to different strains and transgenic toxins of *Bacillus thuringiensis. Environmental entomology, 28*(1), pp.108-115.

Jha KK, McKinley CR. 2002. Scope of agroforestry in Indo-Gangetic plains of India and principles for improvement of selected tree species. ENVIS Forestry Bulletin, 2(2):1-4, 33-43.

Jha KK, McKinley CR. 2006. Tree Improvement: The Seed Orchard Approach. ENVIS Forestry Bulletin, 6(4): 18-25.

Karthick, 2019 Ph.D Thesis, Tamil Nadu Agricultural University, Coimbatore, India.

Kedarnath S. 1974. Genetic improvement of forest tree species in India. Indian Journal of Genetics and Plant Breeding, 34A:367 - 374.

Kha, L.D., 1996. Studies on natural hybrids of Acacia mangium and A. auriculiformis in Vietnam. In *Tree improvement for sustainable tropical forestry. QFRI-IUFRO Conference, Caloundra, 1996.*

Krishnakumar N, 2017. Studies on Biometric, biochemical and molecular characterization of *Santalum album* L. (An endangered species) genetic resources. *Ph.D Thesis, Tamil Nadu Agricultural University, Coimbatore, India.*

Kumar, U., 2002. Methods in Plant tissue Culture. Agrobios (India), Jodhpur. P: 369.

Kumar, V., 2015. Nursery and plantation practices in forestry. Scientific Publishers.

Langridge, W.H. 2000. Edible vaccines. Sci. Am. 283:66–71.

Liang, H., Catranis, C.M., Maynard, C.A. and Powell, W.A., 2002. Enhanced resistance to the poplar fungal pathogen, *Septoria musiva,* in hybrid poplar clones transformed with genes encoding antimicrobial peptides. *Biotechnology letters, 24*(5), pp.383-389.

Litt, M. and Luty, J.A., 1989. A hypervariable microsatellite revealed by in vitro amplification of a dinucleotide repeat within the cardiac muscle actin gene. *American journal of human genetics, 44*(3), p.397.

Lloyd, G. and McCown, B.H. (1981) Commercially-Feasible Micropropagation of Mountain Laurel (*Kalmia latifolia*), by Shoot Tip Culture. Proceeding of the International Plant Propagation Society, 30, 421-427.

Mandal, A.K. and Gibson, G.L. eds., 1998. Forest genetics and tree breeding. CBS Publishers & Distributors.

Marques, J., Abrams, D., Paez, D. and Martinez-Taboada, C., 1998. The role of categorization and in-group norms in judgments of groups and their members. *Journal of Personality and Social Psychology, 75*(4), p.976.

Mullis, K. 1990. Specific enzymatic amplification of DNA in vitro: The polymerase chain reaction. Cold Spring Harbor Symposia on Quantitative Biology, 51(1), pp.263–273.

Murali, K.S., 1993. Differential reproductive success in Cassia fistula in different habitats—A case of pollinator limitation?. *Current Science*, pp.270-272.

Murashige, T; Skoog, F (1962). "A Revised Medium for Rapid Growth and Bio Assays with Tobacco Tissue Cultures". Physiologia Plantarum. 15 (3): 473–497.

Neale, D.B.2007. Genomics to tree breeding and forest health. *Current opinion in Genetics and Development.* 17: 539-544.

Nitsch, J.P. and Nitsch, C. (1969) Haploid Plants from Pollen Grains. Science, 163, 85-87.

Noor Mohammed, 2012, Heterosis Breeding, Seminar Report, TNAU.

Palanikumaran, B. 2015. Inventory, Improvement and Molecular Characterization of *Callophyllum innophyllum* L. and *Pongamia pinnata* L. Genetic Resources. **Ph.D Thesis.** Tamil Nadu Agricultural University, Coimbatore.

Park YS, Bonga JM and Mullin JJ 1995. Clonal Forestry In:Forest Genetics and Tree Improvement (Eds.A.K.Mandal and G.L.Gibson) CBS Publishers and distributors, New Delhi : 143-167 Pages.

Parthiban, K.T., (2001). Seed source variations, molecular characterization and clonal propagation in teak (*Tectona grandis* Linn. F.). **Ph.D.**, Tamil Nadu Agricultural University, Coimbatore, India.

Parthiban, K.T., (2017). Pulpwood Plantation Development in Brazil - A Visit Experience. Eds: Parthiban, K.T and Seenivasan R Plantation and Agroforestry- Pulpwood value chain approach. Scientific Publisher. Jodhpur. P.517 (ISBN: 978-93-86347-98-5).

Parthiban, K.T., Kumar, R.S., Thiyagarajan, P., Subbulakshmi, V., Vennila, S. and Rao, M.G., 2009. Hybrid progenies in Jatropha–a new development. *Current science*, pp.815-823.

Pereira, V.J., Fernandes, D., Carvalho, G., Benoliel, M.J., San Romão, M.V. and Crespo, M.B., 2010. Assessment of the presence and dynamics of fungi in drinking water sources using cultural and molecular methods. *Water Research*, *44*(17), pp.4850-4859.

Phetole Mangena and Phatlane William Mokwala. 2018. Introductory Chapter: Pollination. In: Pollination in Plants: Intechopen Publisher, 1-6.

Phundan singh and Narayanan, 1993, Biometric Techniques in Plant Breeding, kalyani Publishers, Ludhiana. P:187.

Probir Kanti Biswas, Vatsala Piramal, Sudhir,M and Jayanto Achrekar,2014, Agriculture and forest biotechnology, Dominant publisher and distributor, New Delhi.P:571

Punekar, S.A. and Kumaran, K.P.N., 2010. Pollen morphology and pollination ecology of Amorphophallus species from North Western Ghats and Konkan region of India. *Flora-Morphology, Distribution, Functional Ecology of Plants, 205*(5), pp.326-336.

Purohit, S.S., 2004. A Laboratory Manual of Plant Biotechnology Agrobios (India).P: 210

Purohit, S.S., 2010. Agricultural biotechnology. Agrobiois India, New Delhi. P: 1014

Purohit, S.S., 2010. Agricultural biotechnology. Mrs. Sarswati Purohit for student edition, Jodhpur. P:720

Raju, A.J.S. and Rao, M.M., 2016a. Flowering phenology, breeding system, pollinators and fruiting behaviour of *Pavetta tomentosa* (Rubiaceae) Roxb. Ex Sm., a keystone shrub species in the southern eastern ghats forest, Andhra Pradesh, India. *Annali di Botanica, 6*, pp.85-96.

Raju, A.S., Lakshmi, P.V. and Ramana, K.V., 2012. Reproductive ecology of Terminalia pallida Brandis (Combretaceae), an endemic and medicinal tree species of India. *Current Science*, pp.909-917.

Ramawat, K.G. and Merillon, J. M., 1999. Biotechnology: secondary metabolites. Oxford and IBH Publishing co PVT LTD, Newdelhi. P:393

Rao, S.P. and Raju, A.S., 2002. Pollination ecology of the Red Sanders Pterocarpus santalinus (Fabaceae), an endemic and endangered tree species. *Current Science*, pp.1144-1148.

Ravelonandro, M., R. Scorza, J.C.. Bachelier, G. Labonne, L.Levy, V. Damsteegt, A.M. Callahan, and J. Dunez. 1999.Resistance of transgenic *Prunus domestica* to plum poxvirus infection. Plant Dis. 81:1231–1235.

Roberta H. Smith, 1992, Plant Tissue Culture (Techniques and Experiments), Academic Press Inc. London. P: 166.

Sekar, I. 2003. Seed storage, seed source variations, molecular characterization and *in vitro* propagation in *Simarouba glauca* DC. **Ph.D. Thesis**, Tamil Nadu Agricultural University, Coimbatore.

Senthilkumar, R.S., Parthiban, K.T. and Rao, M.G., 2009. Molecular characterization of Jatropha genetic resources through inter-simple sequence repeat (ISSR) markers. *Molecular biology reports, 36*(7), pp.1951-1956.

Sharma, M.V. and Shivanna, K.R., 2011. Pollinators, pollination efficiency and fruiting success in a wild nutmeg, *Myristica dactyloides. Journal of Tropical Ecology, 27*(4), pp.405-412.

Singh B. D. 1994, Plant breeding - principles and methods, Kalyani Publishers, New Delhi. P: 654

Singh N.B., Kumar D., Gupta R, Pundir I. and Tornar, A. 2002. Intraspecific and interspecific hybridization in poplar for production of new clones. ENVIS Forestry Bulletin. 2(2): 11-16

Singh, A., Negi,M.S., Moses, V.K., Venkateswarlu, B., Srivastava, P.S. and lakshmikumaran, M. 2002. Molecular analysis of micropropagated neem plants using ALFP markers for ascertaining clonal fedility. *In Vitro Cell. Dev. Biol. Plant.* 38: 519-524.

Singh, B.D., 2015. Plant breeding: principles and methods. Kalyani publishers.

Singh, P., Singh, P. and Singh, P., 2007. Essentials of plant breeding. Kalyani publishers.

Smulders, M.J.M., Beringen, R., Volosyanchuk, R. et al. Tree Genetics & Genomes (2008) 4: 663

Soehartono, T. and Newton, A.C., 2001. Reproductive ecology of Aquilaria spp. in Indonesia. *Forest ecology and management, 152*(1-3), pp.59-71.

Soto, A., Lorenzo, Z. and Gil, L., 2003. Nuclear microsatellite markers for the identification of Quercus ilex L. and Q. suber L. hybrids. Silvae Genetica, 52(2), pp.63-66.

Stokes, V., Lee, S., Forster, J. and Fletcher, A., 2018. A comparison of Sitka spruce x white spruce hybrid families as an alternative to pure Sitka spruce plantations in upland Britain. *Forestry: An International Journal of Forest Research, 91*(5), pp.650-661.

Surendran, C., Parthiban, K.T., Vanangamudi, K. and Balaji, S., 2000. Vegetative propagation of trees, principles and practices. Coimbatore, India.

Thirumurugan, D., Cholarajan, A., Raja, S.S. and Vijayakumar, R., 2018. An Introductory Chapter: Secondary Metabolites. *Secondary Metabolites: Sources and Applications,* p.1-21.

Thiruniraiselvan, R, 2018. Improvement and Utilization of *Neolamarckia cadamba* Roxb. Genetic Resources for Feed Quality and Utility. **Ph.D thesis.** Tamil Nadu Agricultural University, Coimbatore.

Vanangamudi, K., Bharathi, A., Natesan, P., Jerlin, R., Thiyagarajan, T. M., and Kanniyan, S.,2001, Recent techniques and participatory Approaches on Quality Seed production, Tamil Nadu Agricultural University, Coimbatore.

Vanangamudi, K., Paramathma, M., Parthiban, K.T., Natarajan, K., Sathyasheela, K. R. V., and Saravanan, T., 2006. Vegetative propagation of trees, principles and practices, International book Distributors

Venkatesh, C.S. and Sharma, V.K., 1976. Heterosis in the flowering precocity of Eucalyptus hybrids. *Silvae Genetica, 25*(1), pp.28-29.

Vennila, S. 2009. Pulpwood traits, genetic and molecular characterization of *Eucalyptus* genetic resources. **Ph.D thesis.** Tamil Nadu Agricultural University, Coimbatore.

Verma, K.S., Khurana, D.K. and Christersson, L., 2006. Short rotation forestry for industrial and rural development. In IUFRO-ISTS-UHF International Conference on World Perspective on Short Rotation Forestry for Industrial and Rural Development (2003: Solan, India). Published for Indian Society of Tree Scientists, Solan by Westville Pub. House.

Xu, Y.2010. Molecular Plant Breeding. *CABI, Wallingford,* UK. Pp. 27.

Zobel B and Talbert J 1984 Applied Forest Tree Improvement, John Wiley and Sons, New York: 505 Pages.

